T0233940

In-/Near-Memory Computing

Synthesis Lectures on Computer Architecture

Editor
Natalie Enright Jerger, *University of Toronto*

Editor Emerita
Margaret Martonosi, *Princeton University*

Founding Editor Emeritus
Mark D. Hill, *University of Wisconsin, Madison*

Synthesis Lectures on Computer Architecture publishes 50- to 100-page books on topics pertaining to the science and art of designing, analyzing, selecting, and interconnecting hardware components to create computers that meet functional, performance, and cost goals. The scope will largely follow the purview of premier computer architecture conferences, such as ISCA, HPCA, MICRO, and ASPLOS.

In-/Near-Memory Computing
Daichi Fujiki, Xiaowei Wang, Arun Subramaniyan, and Reetuparna Das
2021

Robotic Computing on FPGAs
Shaoshan Liu, Zishen Wan, Bo Yu, and Yu Wang
2021

AI for Computer Architecture: Principles, Practice, and Prospects
Lizhong Chen, Drew Penney, and Daniel Jiménez
2020

Deep Learning Systems: Algorithms, Compilers, and Processors for Large-Scale Production
Andres Rodriguez
2020

Parallel Processing, 1980 to 2020
Robert Kuhn and David Padua
2020

In-/Near-Memory Computing

Synthesis Lectures on Computer Architecture

Editor
Natalie Enright Jerger, *University of Toronto*

Editor Emerita
Margaret Martonosi, *Princeton University*

Founding Editor Emeritus
Mark D. Hill, *University of Wisconsin, Madison*

Synthesis Lectures on Computer Architecture publishes 50- to 100-page books on topics pertaining to the science and art of designing, analyzing, selecting, and interconnecting hardware components to create computers that meet functional, performance, and cost goals. The scope will largely follow the purview of premier computer architecture conferences, such as ISCA, HPCA, MICRO, and ASPLOS.

In-/Near-Memory Computing

Daichi Fujiki, Xiaowei Wang, Arun Subramaniyan, and Reetuparna Das

ISBN: 978-3-031-00644-9 paperback
ISBN: 978-3-031-01772-8 ebook
ISBN: 978-3-031-00069-0 hardcover

DOI 10.1007/978-3-031-01772-8

A Publication in the Springer series
SYNTHESIS LECTURES ON ADVANCES IN AUTOMOTIVE TECHNOLOGY

Lecture #57
Series Editor: Natalie Enright Jerger, *University of Toronto*
Editor Emerita: Margaret Martonosi, *Princeton University*
Founding Editor Emeritus: Mark D. Hill, *University of Wisconsin, Madison*
Series ISSN
Print 1935-3235   Electronic 1935-3243

In-/Near-Memory Computing

Daichi Fujiki, Xiaowei Wang, Arun Subramaniyan, and Reetuparna Das
University of Michigan, Ann Arbor

SYNTHESIS LECTURES ON COMPUTER ARCHITECTURE #57

ABSTRACT

This book provides a structured introduction of the key concepts and techniques that enable in-/near-memory computing. For decades, processing-in-memory or near-memory computing has been attracting growing interest due to its potential to break the memory wall. Near-memory computing moves compute logic near the memory, and thereby reduces data movement. Recent work has also shown that certain memories can morph themselves into compute units by exploiting the physical properties of the memory cells, enabling in-situ computing in the memory array. While in- and near-memory computing can circumvent overheads related to data movement, it comes at the cost of restricted flexibility of data representation and computation, design challenges of compute capable memories, and difficulty in system and software integration. Therefore, wide deployment of in-/near-memory computing cannot be accomplished without techniques that enable efficient mapping of data-intensive applications to such devices, without sacrificing accuracy or increasing hardware costs excessively. This book describes various memory substrates amenable to in- and near-memory computing, architectural approaches for designing efficient and reliable computing devices, and opportunities for in-/near-memory acceleration of different classes of applications.

KEYWORDS

processing-in-memory, near-memory computing, in-memory computing, SRAM, DRAM, non-volatile memories, ReRAM, memristor, flash memory, accelerator architecture, domain-specific accelerators

Contents

Preface

General purpose processors and accelerators, including graphics processing units, follow the Von Neumann architecture. It is composed of three components: processor, memory, and interconnection of these two. This simple but powerful model has been the basis of computer architecture since the very first computer was born, while the recent data-intensive trend in computation workloads has seen bottlenecks in this fundamental paradigm of computers. Studies show data communication takes 1,000× time and 40× power compared to arithmetic done in the processors. This is the "memory wall."

For decades, Processing-in-Memory (PIM) has been an attractive idea that has the potential to break the memory wall. PIM moves compute logic near the memory and thereby reduces data movement. The advent of commercially feasible 3D stacking technologies, such as Micron's Hybrid Memory Cube (HMC), has renewed interest in PIM. Furthermore, certain memories have been shown that they can morph themselves into compute units by exploiting the physical properties of the memory cells and arrays. Enabling in-place computing in memory without transferring data outside memory arrays frequently unlocks massive parallelism and computation horsepower due to dense memory arrays while significantly reducing the data communication cost.

In order to understand the basics and the state of the art of in-/near-memory computing, this book aims to provide an overview of various memory technologies, the key concepts that enable in- and near-memory computing in each memory substrate, and the techniques that have been explored to improve the performance and broaden the applicability. We hope this book will serve as a useful introduction to in-/near-memory computing for readers who have a basic understanding of computer architecture but are new to this field. We also hope this book will help develop new ideas to expand the horizon of future compute-capable-memory technologies.

ORGANIZATION

This book is organized as follows.

- Chapter 1 provides a brief introduction and the background of in-/near-memory computing.

- Chapter 2 is intended to clarify the taxonomy of various in-/near-memory computing and to make a comparison of the salient features of each of the classes of the memory-centric computing approaches.

- Chapter 3 provides a historical perspective of early PIM works and deep dives into 3D PIM architectures.

- Chapter 4 describes in-memory computing in SRAM enabled by digital and mixed-signal approaches. It highlights the bitline computing technique, one of the representative methods to accomplish in-memory computing.

- Chapter 5 describes efforts to re-purpose emerging Non-Volatile Memories (NVMs) for computation. Some NVMs have desirable cell properties that can be exploited for in-memory computation in the analog domain. This chapter introduces various efforts to enhance the efficiency and applicability of such computing schemes. It also explains near-storage computing in solid-state drives and bulk bit-wise computing in digital NVMs.

- Chapter 6 provides an overview of novel accelerator architectures using in-memory and near-memory computing. While machine learning has taken the leading part of the memory-centric acceleration, we also cover other promising applications that can benefit from this concept, including automata processing graphs, database, and genomics, with their unique challenges for in-/near-memory computing.

- Chapter 7 describes programming and system support for in-/near-memory computing.

Each chapter has a small pitfall section that describes popular misconceptions of in-/near-memory techniques covered in the chapter.

Daichi Fujiki, Xiaowei Wang, Arun Subramaniyan, and Reetuparna Das
June 2021

Acknowledgments

The authors are grateful to Natalie Enright Jerger (current editor) and Michael Morgan (President and CEO of Morgan & Claypool Publishers) for their support throughout the entire process of preparing this book. This book was written over the course of the COVID pandemic, and would have been impossible without their patience and encouragement. Many thanks to our collaborators for in-memory computing research: David Blaauw, Satish Narayanasamy, Dennis Sylvester, Ravi Iyer, Supreet Jeloka, Shaizeen Aga, Jingcheng Wang, Scott Mahlke, Westley Weimer, Kevin Skadron, Kyojin Choo, Charles Eckert, and Charles Augustine. The book draws upon years of research with these collaborators, and countless discussions. We would like to thank Onur Mutlu and Rajeev Balasbrominan for several technical discussions on the topics covered in this book. We also greatly appreciate the feedback of anonymous reviewers of this book. The authors are grateful to research sponsorship from Nakajima foundation fellowship, Intel, National Science Foundation, ADA center under JUMP-SRC program, and CFAR center. Finally, the authors deeply appreciate their family members for their unconditional love and support.

Daichi Fujiki, Xiaowei Wang, Arun Subramaniyan, and Reetuparna Das
June 2021

C H A P T E R   1

Introduction

Computer designers have traditionally separated the roles of storage and computation. Memories stored data. Processors computed them. Is this distinction necessary? A human brain does not separate the two so distinctly, so why should a computer? [1] Before addressing this question, let us start with the well-known memory wall problem. What is the memory wall in today's context?

The memory wall [2] originally referred to the problem of growing disparity in speed between fast processors and slow memories. In the last three decades, the number of processor cores per chip has steadily increased while memory latency has remained relatively constant. This has led to the so-called memory wall where memory bandwidth and memory energy have come to dominate computation bandwidth and energy. With the advent of data-intensive applications, this problem is further exacerbated. Today, a large fraction of energy is spent in moving data back-and-forth between memory and compute units. For well over three decades, architects have tried a variety of strategies to overcome the memory wall. Most of them have centered on exploiting locality and building deeper memory hierarchies.

Another alternative: what if we could move computation closer to memory—so much that the line that divides computation and memory starts to blur? The key idea is to physically bring the computation and memory units closer together as a solution to scale the memory wall. This approach has several flavors—ranging from placing computation units near the memory banks, to even more exciting technologies that dissolve the line that distinguishes memory from computational units. There are two main benefits of merging compute and memory. It eliminates data movement and hence significantly improves energy efficiency and performance. Furthermore, it takes advantage of the fact that over 90% of silicon in today's compute platforms simply stores and provides data retrieval; harnessing this area by re-purposing it to perform computation can lead to massively parallel computational processing. Chapter 2 attempts to create a taxonomy of the diverse landscape of memory-centric processing, while discussing technology basics and related trade-offs.

Early works in this field placed discrete compute units inside the main memory (DRAM), an approach popularly referred to as processing in memory (PIM). Researchers discussed PIM in the 1990s [3–7] (initial suggestions date back to as early as the 1970s [8]). But this idea did not quite take off back then, due to the high cost of integrating computational units within a DRAM die. Another factor may have been the fact that cheaper optimizations were still possible, thanks to Moore's law and Dennard scaling. The advent of commercially feasible 3D chip stacking technologies, such as Micron's Hybrid Memory Cube (HMC), renewed interest in PIM in the early 2010s. HMC stacks layers of DRAM memory on top of a logic layer. Computational units in the logic layer can communicate with memory through high-bandwidth through-silicon vias. Thanks to 3D integration technology, we can now take computational and DRAM dies implemented in different process technologies and stack them on top of each other. The additional dimension in 3D PIM allows an order of magnitude more physical connections between the computational and memory units, and thereby provides massive memory bandwidth to the computational units. Chapter 3 provides a historical perspective of early PIM works and deep dives into 3D PIM architectures.

Although PIM brings computational and memory units closer together, the functionality and design of memory units remain unchanged. An even more exciting technology is one that dissolves the line that distinguishes memory from computational units. Over 90% of silicon in processor and main memory dies is simply to store and access data. What if we could take this memory silicon and re-purpose it to do computation? The biggest advantage of this approach is that memory arrays morph into massive vector computing units (potentially, one or two orders of magnitude larger than a graphics processing unit's (GPU's) vector units), as data stored across hundreds of memory arrays could be operated on concurrently. Because we do not have to move data in and out of memory, the architecture naturally saves the energy spent in those activities, and memory bandwidth becomes a meaningless metric.

Re-purposing memory for computing has emerged as a mainstream technology over the last decade. For instance, consider Micron's Automata Processor (AP) [9] as an example. It transforms DRAM structures to a Nondeterministic Finite Automata (NFA) computational unit. NFA processing occurs in two phases: state match and state transition. AP cleverly re-purposes the DRAM array decode logic to enable state matches. Each of the several hundreds of memory arrays can now perform state matches in parallel. The state-match logic is coupled with a custom interconnect to enable state transition. We can process as many as 1,053 regular expressions in Snort (a classic network-intrusion detection system) in one go using little more than DRAM hardware. AP can be an order of magnitude more efficient than GPUs and nearly two orders of magnitude more efficient than general-purpose multicore CPUs!

AP re-purposed just the decode logic in DRAMs. Another line of research demonstrates that it is possible to re-purpose SRAM array bitlines and sense-amplifiers to perform in-place computation on the data stored in SRAM [10, 11]. Data stored in memory arrays share wires (bitlines) and signal sensing apparatus (sense-amps). We observe that logic operations can be computed over these shared structures, which allows us to re-purpose thousands of cache memory arrays into over a million of bit-serial arithmetic-logic units. Thus, we morph existing cache into massive vector compute units, providing parallelism several orders of magnitude higher than a contemporary GPU. This style of bitline computing may work well in a diverse set of memory technologies (DRAMs, RRAMs, STT-MRAMs, and Flash). Further, the degree of analog computing can be tuned and leads to a diverse design space. Chapters 3, 4, and 5 are dedicated to covering the foundations of this emerging technology, breaking it down by specific underlying memory bit-cell type (DRAM, SRAM, and Non-Volatile Memories).

Memory is central to all computing devices, both general-purpose processors as well as accelerators. Latest Intel's server-class Xeon processor, for instance, devotes 35 MB just for its last-level cache, while Google's Tensor Processing Unit [12] dedicates several tens of MBs for on-chip storage. Using memory for computation is thus a feasible and profitable approach for both general-purpose processors, and domain-specific accelerators. Chapter 6 provides an overview of these novel accelerator architectures while summarizing the key insights which drive In-Memory and Near-Memory accelerators.

Finally, compute capable memories require a programming model and compiler that is capable of exposing parallelism/data-movement in applications to the underlying hardware and harnessing its full potential. Chapter 7 provides a summary of the various programming models used, and highlights the decision drivers for the choice of the programming models.

CHAPTER 2

Technology Basics and Taxonomy

The terms near memory computing and in-memory computing are sometimes used interchangeably and confusingly. This chapter aims to clarify the taxonomy of various near and in-memory computing approaches and make a comparison of the salient features of each class of memory-driven approaches. In addition, a computable memory device can be implemented as a discrete accelerator device or as a memory module that replaces the one in the current memory hierarchy. We will explore the benefits and challenges posed by each of the approaches.

2.1 IN VS. NEAR MEMORY

In this section, we will present the taxonomy of various in and near memory computing approaches. The boundary of in and near memory computing is sometimes fuzzy. Moreover, they can be used in a different context (such as an in-memory database). In this book, we focus on prior research efforts that change the memory architecture for computation, their computation schemes, data access schemes, and the proximity of data processing to the memory to classify them. Our classification (see Figure 2.1) adopts the insights of [13], which suggests a classification based on where the computation result is produced.

2.1.1 PROCESSING IN MEMORY AND NEAR MEMORY COMPUTING

Near memory computing literally performs the computation in logic placed near the memory. Near memory computing architectures were first termed processing in memory (PIM).

Breaking the memory wall has been the primary goal of these memory-centric architectures. Since the 1990s (initial suggestions are dated back to as early as the 1970s), PIM has attracted researchers' attention as an alternative approach to overcome the memory bandwidth limitation of the Von Neumann architecture. The key idea was to physically couple the computation and memory units close together by placing computation units inside the main memory (DRAM). This class of classical PIM approaches will be further explained in Section 3.1.

The traditional PIM approaches face several critical challenges in integrating computational units within a DRAM die. However, since the 2010s, the advent of commercially available 3D stacked memories has created a renewed interest in PIM. For example, Micron's Hybrid Memory Cube (HMC) incorporates a logic layer under a stack of DRAM layers, advocating

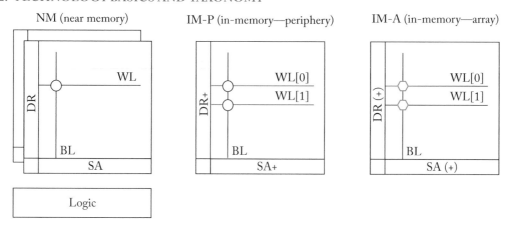

Figure 2.1: Taxonomy of in and near memory computing. Computation happens in modules colored in yellow. SA = Sense Amplifier; DR = Wordline Driver; WL = Wordline; BL = Bitline.

promisingly that a custom logic can be implemented within the logic layer. PIM in the context of 3D stacked memory is covered in Section 3.2.

PIM is frequently referred to as *near memory computing* these days to avoid confusion with in-memory computing, a new paradigm of memory-centric computing, which we define in the following section. The critical distinctions of near memory architectures from traditional Von Neumann architectures are the following.

1. Compute logic is placed close to the memory, often using high bandwidth circuit integration techniques (e.g., 2.5D and 3D integration) to leverage large memory access bandwidth available internally in the memory.

2. Memory cells, memory arrays, and peripheral circuits that provide fundamental read and write access to data in the memory cells are typically kept intact.

2.5D integration circuit adopts a silicon interposer or an organic interposer to connect a memory die and a logic die, enabling high wiring density and power efficiency compared to the traditional wire bonding on a printed circuit board (PCB). 3D integration uses interlayer connection techniques such as Through-Silicon Via (TSV) and microbump to stack DRAM layers. Both contribute to providing a large internal memory bandwidth and technology friendliness as the logic die can use a different process technology optimized for logic, motivating PIM in stacked memories. Moreover, the fundamental architecture and protocol for accessing memory cells are unchanged. Thus, it saves the enormous design cost of constructing an entirely new memory device. Some near memory computing devices are already commercially available for these reasons [14–16].

Pitfall A general-purpose core can be implemented for PIM to provide flexible processing.

This is not true because of the following reasons.

- The available memory bandwidth is so high in these systems that a general-purpose multicore processor with tens of cores is a poor candidate to take advantage of 3D PIM.

- Many applications written in imperative programming languages exploit temporal and spatial locality, which let them reap substantial benefits from the cache hierarchy. PIM rarely has such a cache structure. The wide memory bandwidth of PIM is better exploited by a class of applications that can expose parallelism or need large bandwidth.

- Area that can be allocated for logic is small in PIM compared to a CPU die. The cost for logic in PIM is typically worse than general logic die due to the difference in process technology.

- The thermal dissipation requirement is often challenging for a general core.

2.1.2   IN-MEMORY COMPUTING

In-memory computing is a new paradigm of memory-centric computing, which inherits the spirits of PIM and near memory computing. While near-memory computing implements a logic circuit independently from the memory structure, in-memory computing intimately involves memory cells, memory arrays, and peripheral circuits in computation. Structural modifications of them or additional special circuitry are often required to support computation.

Historically, in-memory computing was considered an economically infeasible design. Modifying memory cells adds non-trivial reinvestment cost to the memory design whose technology is deeply optimized to a uniform cell structure of the current memory architecture. Moreover, the outcome of the modified cell design would drastically lower the density, which could make it challenging for a memory-centric architecture to justify its performance vs. area (or performance vs. cost) trade-off.

With the emergence of non-volatile memories (NVMs), the idea of in-memory computing has been revisited. Certain NVMs have a desirable physical property to perform computation in the analog domain, enabling in-memory computing with minimal design changes to the memory arrays. Moreover, the non-volatile nature of the memory cells solves the issues of disruptive read access of DRAM cells, which has forced in-DRAM computing to perform a copy before computing. On the other hand, in-memory computation in the analog domain is still a speculative technology. For example, the non-ideality that exists due to process variation and extended current paths may compromise the computation results. Moreover, digital to analog conversion (DAC) and analog to digital conversion (ADC) costs will get prohibitive as more bits are handled by the conversion of the analog signals.

Along with this stream, researchers revisited in-memory computing designed for present memory substrates, i.e., SRAM, DRAM, and NAND flash memory. They address the challenges above and leverage the matured technology of these memories. Some research works propose digitalized computation in NVMs for reliability. We will show the DRAM-based in-memory computing in Chapter 3, SRAM-based in-memory computing in Chapter 4, and NVM-based in-memory computing in Chapter 5.

In-memory computing approaches can be further subdivided into two classes, in-memory (array) and in-memory (periphery).

- In-memory (array) or **IM-A** produces the computing results within the memory array using a special compute operation (e.g., MAGIC [17] and Imply [18], explained in Chapter 5) for computation. IM-A architectures can offer maximum bandwidth and energy efficiency because operations happen inside the memory array. IM-A can also provide the largest throughput for simple operations. On the other hand, complex functions may incur high latency. Moreover, IM-A often requires redesigning the memory cells for such special compute operations, extending the normal bit- and word-line structures. Since the design and layout of the cells and arrays are heavily optimized for specific voltages and currents, any changes in the cell and array access method incur heavy redesigning and characterization efforts. Furthermore, modification to the peripheral circuit (i.e., the logic circuit required to perform read and write operations, such as word-line drivers and sense-amplifiers) is sometimes required to support IM-A computation. Therefore, IM-A includes (a) IM-A with major changes in the memory arrays, and (b) IM-A with major changes in the memory arrays and minor changes in the peripheral circuit.

- In-memory (periphery) or **IM-P** produces the computing results within the peripheral circuits. IM-P can be further divided into digital IM-P approaches, which only handle digital signals, and analog IM-P or **IM-P (analog)** approaches, which perform computation in the analog domain. The modified peripheral circuits enable operations beyond the normal read/write, such as interacting with different cells or weighting read voltages. Such modifications include support for multiple row activation in the wordline driver and DACs/ADCs for multi-level activation and sensing. They are designed for computation ranging from logic operations to arithmetic operations such as dot-product in vector-matrix multiplication. Although the results are produced in the peripheral circuits, the memory array performs a substantial amount of computation. The changes to the peripheral circuit may necessitate currents/voltages for arrays that

Table 2.1: Comparison of conventional Von Neumann architectures (baseline) with near-memory computing (NM), IM-A, and IM-P (digital and analog) architectures

		Baseline	NM	IM-P	IM-P (Analog)	IM-A
Cell modication		No	No	No	No	Yes
Peripheral modication		No	No	Yes	Yes	Yes
Density	Cell	High	High	High-Mid	High Low	Low
	Overall		Low-Mid			
Memory ↔ compute unit bandwidth		Low	Mid	High	High	High
Energy		High	Mid	Low	Low	Low
Area for logic		High	Mid	Low-Mid	Low-Mid	Low
Dataflow flexibility		High	Mid	Low	Low	Low
Logic flexibility		High	Mid	Low	Low	Low
Computation latency		Low	Low	Mid-High	Mid	High
Precision flexibility		High	High	Mid-High	Low-Mid	Low
Reliability, ECC support		High	High	Mid	Low	Mid

are different from those used in conventional memories. Thus, IM-P may use a slightly different cell design for robustness. Additional circuits to the peripheral for supporting complex functions may lead to high costs.

In-memory computing enables extensively large bandwidth access to the data compared to the near-memory approaches. It is exploited not only for addressing the memory wall problem but also for massively parallel processing.

2.1.3   COMPARISON OF IN AND NEAR MEMORY COMPUTING

We provide a comparison of the conventional Von Neumann architectures (baseline), near-memory computing (NM), IM-A, and IM-P architectures in Table 2.1.

- **Modification of cells and peripheral circuits**: The baseline and NM architectures use the memory system as-is. Thus, no modification is required. IM-P modifies peripheral circuits for special compute operations, and IM-A may require cell modifications.

- **Density**: Since the memory arrays are heavily optimized, cell density is the highest when the memory array macro is used as-is. Importantly, overall density (array + peripheral) is less sensitive to on-die logic when a logic-friendly memory substrate (e.g., SRAM, eDRAM) or advanced integration techniques such as 3D stacking are used.

Some of the classical NM architectures implement logic in the same DRAM die using the DRAM process technology. Such a design may significantly lessen the overall memory density. IM-P can face the same issue as NM, but oftentimes a smaller amount of change is required than NM. This is because a substantial part of computation happens in the memory arrays, requiring smaller additions in the peripheral to implement the same processing element as NMs; thus, the density can be less affected. IM-P (analog) has a higher density for cell storage, but it often comes at the cost of a more significant area requirement for peripheral if ADCs are needed.

- **Bandwidth between memory and computation unit**: Memory bandwidth is reduced when the computation unit is placed far from the memory. Also, the computation unit needs to support extensive parallelism to keep up with the large bandwidth. Thus, the computation bandwidth requirement is highly related to the memory bandwidth.

- **Energy**: Data delivery consumes a significant amount of energy compared to the energy for computation. In general, if the data travels less, less energy is consumed.

- **Area**: Area can be interpreted as (a) logic area required to do one arithmetic operation (e.g., addition), and (b) die area available for logic implementation. Baseline and NM require standard area for logic (a) but can afford a large die area (b) and a flexible logic implementation. IMs re-purpose memory arrays for computation, thus requiring less area for logic (a), but its available die area (b) is limited.

- **Dataflow flexibility**: Some applications require non-uniform memory access such as random access and indirect access (e.g., mem[addr[i]]). In order to accomplish such irregular access, compute units need to have global accessibility to memory contents. NM and IMs may have access to a limited region of the memory address space, and remote access incurs expensive all-to-all communication between memory nodes or memory arrays.

- **Logic flexibility**: The area budget for logic puts a limitation on the logic complexity that can be implemented. IM-A typically has only several extra diodes per cell, and IM-P has some dozens of gates per bitline. IMs use a combination of basic operations or resort to an external processing unit to supplement operations.

- **Computation latency**: Because of the logic complexity constraints, IMs often perform iterative operations to perform one arithmetic operation, resulting in a larger computation latency. On the other hand, IMs usually have a large computation bandwidth which can compensate for the latency.

- **Precision flexibility**: The baseline and NM architecture can implement arithmetic logic with any precision, including floating points. Digital IM approaches can combine several bit operations to compose logic with arbitrary precision. It usually requires interaction of multiple bit-cells across rows or columns, so they are typically classified into IM-P. IM-P (analog) can operate at higher bit precision per bitline. While the bit precision for analog computation is limited by many circuit factors, e.g., capacitance and ADC resolution, multiple results can be composed to produce arbitrary integer precision. However, it is challenging to extend it for floating point precision.

- **Reliability and ECC support**: Memories are susceptible to various error sources such as hard errors (e.g., cell failure) and soft errors (e.g., bit flips due to cosmic radiation). Memories protect themselves from such errors using error correcting codes (ECCs), but we have few works on ECC that are compatible with in-memory computing. Moreover, computation in an analog domain leads to the addition of analog noises. Some analog IM-P architectures use a small number of bits per cell to increase the noise margin, or use an aggressive (error-prone) cell configuration for error-tolerant workloads such as machine learning for which the model can be trained to tolerate such errors and noises.

We have shown in-/near-memory computing has different interesting trade-offs. In the following chapters, we will present representative works of each class of architectures and discuss what kind of parallelism is exploited, what kind of application is suitable, and how the programming model and execution model can leverage the parallelism and their computation horsepower.

2.2   DISCRETE ACCELERATORS VS. INTEGRATED IN-MEMORY HIERARCHY

Memory-centric architectures combine the roles of memory and computation in a memory module. That being said, an NM or IM memory module can be either designed as a discrete accelerator or as an accelerator integrated into a memory module in an existing memory hierarchy, as shown in Figure 2.2.

A discrete accelerator has full access to its memory space without constraints, similar to a scratchpad memory. A discrete memory space decouples an accelerator from operating system paging policies, coherence protocols, data scrambling, and address scrambling. It also provides a control for flexible data arrangement. In particular, most IM architectures require aligning operands within specific columns of a specific array or transposing the input to process it in a bit-serial manner. Discrete accelerators can support these architecture-specific data layouts without much complication. The user interface can be provided as library function calls linked with their driver, similar to an ASIC accelerator. One of the important shortcomings of discrete accelerators is that they still require to load data from the memory hierarchy through an external link such as PCIe, which is likely to be a bottleneck. This issue persists in commodity accelerators as well: a GPU takes non-trivial time copying data to and from the host memory via PCIe bus. This data loading cost can be amortized by reusing the data over time. Thus, the applications

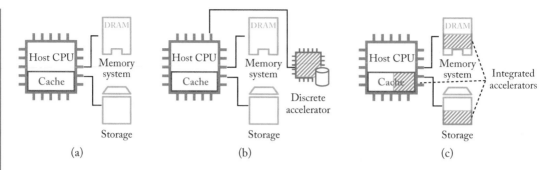

Figure 2.2: (a) Baseline system. (b) A system with a discrete accelerator. (c) A system with an accelerator integrated into its memory hierarchy.

that can achieve high performance are often limited to those that present high reuse or high GOPs (giga-ops) per byte.

An integrated accelerator is ideal for bypassing the memory wall. However, many existing schemes and constraints in each layer of the memory hierarchy, which are implemented for access performance and security, make it challenging to design a full-fledged integrated NM/IM system. For example, in order to align operands in an SRAM subarray prior to computation, assigning adequate addresses to them is not sufficient; they need to be associated with a specific way. DRAM uses various scrambling techniques, and also obtaining a virtual address for operand access needs to go through the OS's page table. NAND flash uses a flash translation layer (FTL), which adds another layer of address translation and is encapsulated in the flash device. Many NVMs have limited write endurance, and these translation layers are helpful for wear-leveling. Interference in them can end up shortening the memory cell lifetime. An integrated system needs to get along with these existing frameworks, including OS and programming models, but we do not have a complete solution yet.

A discrete accelerator and an integrated accelerator are not mutually exclusive options. There can be a hybrid approach of both. For example, we can create a scratchpad memory in an existing memory hierarchy. It still needs to copy data from the memory of the same or lower hierarchy, but compared to importing and exporting data using a shared bus (e.g., PCIe), one can expect higher memory bandwidth. The driver can also free the designated scratchpad memory anytime after computation so that it can be utilized as a standard memory space.

C H A P T E R   3

Computing with DRAMs

3.1 HISTORY, OLDER ARCHITECTURES, AND TECHNOLOGY EVOLUTION

In modern computer architecture, the main memory is usually built out of DRAM chips. The processor is on a separate chip and when the processor needs data from memory, the data is transferred on the interconnect between the processor and memory chips. The interconnect is usually implemented as data buses with restricted bandwidth, thus limiting the data transfer rate. At the same time, the speed of the processors has been improving steadily. Therefore, the data transfer between processor and memory has become the performance bottleneck of the applications that access a large amount of memory data.

Thus, a faster way of moving data between the DRAM and the processor may improve the system performance. The early explorations of near-DRAM processing (1970s–1990s) were inspired by this idea. A common solution is to integrate the processor and the DRAM onto the same chip, replacing the inter-chip communication with on-chip data movement. Figure 3.1 shows a typical system structure with processor and DRAM integrated together. The near-DRAM processing system mainly consists of the scalar processing system, the memory system and vector processing units. The scalar processing system still has a similar structure as in a CPU, with the processor core and the cache system. The memory interface unit interacts with the cache system and the DRAM modules, serving as a bridge enabling near-DRAM processing. To fully take advantage of the high bandwidth of the integrated DRAM, vector processing units are placed to perform data-parallel operations. The vector processing unit has its own vector registers and vector arithmetic units, and interacts with the memory interface unit for accessing memory data. As the vector registers have a wide data width, vector processing benefits naturally from the increased memory bandwidth. The memory system itself is organized in multiple DRAM modules. It is designed with wider I/O interface than a normal DRAM chip to obtain a higher aggregated data bandwidth to serve the processing systems. At runtime, the processors function in a similar way as in a normal system, except that when data are needed from memory, they are directly transferred through the on-chip interface.

Multiple advantages arise from the above NMP system. First, the memory bandwidth has significantly increased, as the external memory data bus is no more the bottleneck. Second, the latency for memory access is also shorter, as the long wire delay in the external data bus can be eliminated. Finally, the system energy efficiency is also improved, as the on-chip data movement consumes less energy than the off-chip data bus.

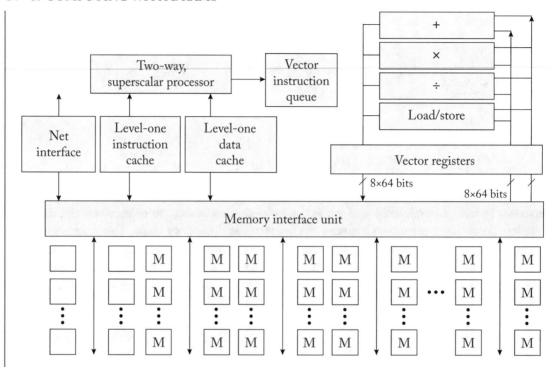

Figure 3.1: Earlier NMP architecture (Intelligent RAM) [19].

The major disadvantage of the above NMP design is that such an integrated processor has a weaker performance than the normal processor in a standalone chip. This is because the processor is implemented with the DRAM process to be integrated in the same chip. The logic circuit and the SRAM arrays of the processor are all slower when implemented in the DRAM process, which is optimized for memory cost and energy, but not logic speed. Further, the in-DRAM logic only has access to the 2–3 metal layers used in the DRAM process; even simple logic circuits will have larger than usual footprints.

Many variations on the above NMP design have been explored in literature. For instance, simple in-order and customized processing elements have been proposed for the integrated-processor NMP system. These NMP designs can also be categorized based on the proximity of the processing elements to the DRAM arrays. Below we discuss both these design alternatives in more detail.

Variations Based on Processing Element Complexity   There are several alternative designs with varied processors in the above NMP system. First, the integrated processor systems can be replaced with simpler processors (for example, in-order pipelined cores) to save area and reduce design complexity [20]. Second, for application-specific designs, the processing elements with customized structures can be used instead of the general purpose scalar and vector processors. For example, the cache system can be replaced with application specific data buffers, and the ALU can be optimized for the specific operation set [21].

Variations Based on Proximity of Processing Elements to Memory Cells   Another series of variations is to move the processing elements closer to the memory cells [22]. The main design described above has separate processors and memory units. To have a tighter integration of the processing elements with the memory, there are designs to put the processing elements near the individual DRAM arrays, or even closely coupled with every single memory cell. As the process elements must have a limited size and complexity to be distributed more densely, they are usually designed for specific applications.

Processing elements near DRAM arrays: Simpler processing elements can be integrated into the peripherals of the DRAM arrays. For example, a PE may consist of several registers and a simple reconfigurable lookup table [22]. The inputs to the lookup table are a few bits both from the DRAM array and the registers. The PE results can be directly written back to the DRAM arrays, or stored in the registers for future operations. To process the data with large width, a PE can iteratively process different parts of data. Throughout the computation, the host CPU controls the execution by sending instructions to the PEs.

Processing elements coupled with DRAM cells: The processing elements can also be integrated as part of the memory cell in the DRAM arrays [8, 23–25]. The processing element here usually only contains a few logic gates to avoid excessive area cost. The logic gates are designed according to the application requirements. To enable co-processing on data in different cells of the array, an in-array data transfer system is deployed, where each cell may transfer its data value to its neighboring cells (for reducing the interconnect complexity). By performing a series of single-cell computation and in-array data movement, an operation is done collectively by all cells in the array. If the application requires more complex processing, additional logic such as arithmetic units can be placed outside of arrays and assist the computation [8].

In summary, the PE-memory mixed arrays can achieve high data parallelism and ultra-short distance between PE and data storage. However, they lack computation generality given the limited number of logic gates and difficulties in intra-array data movement. The customized memory cells also result in high design complexity. More importantly, they suffer from the reduced logic performance when implemented in a DRAM process as well.

3.2   3D STACKED DRAM AND 3D NMP ARCHITECTURES

The near-DRAM processing designs in the 1990s did not flourish, mainly because of the difficulties in integrating the cost/power-aware DRAM and performance-aware computing logic on the same die. As the need for higher memory bandwidth grows with modern data-intensive applications such as big data analytics, the 3D stacked memory has evolved as a novel technology to satisfy such needs.

3.2.1   3D STACKED DRAM

A 3D stacked memory is a 3D integrated circuit with multiple heterogeneous 2D dies stacked on each other. Each 2D die becomes a layer in the 3D stack. The majority of the layers are memory layers, where the dies are built out of DRAM modules. The layer in the bottom of the stack mainly consists of controlling logic, and thus is known as the logic layer. Vertically, the dies communicate with each other with through-silicon vias (TSV) and micro-bumps. By such vertical communication, the travel distance of data is overall reduced, thus improving the latency and energy efficiency.

With DRAM and logic built on separate dies, they may use distinct transistor technology. Thus, logic layers can continue to achieve high performance while memory layers optimize for cost and power efficiency. This marks a key advantage of 3D stacked DRAM for NMP: in previous works where the computing logic and memory are integrated on the same die, the logic needs to be built with DRAM technology and suffers from reduced performance.

In 3D stacked memory, all the performance-critical DRAM control logic is moved to the logic layer. The logic layer is responsible for generating the commands to access the DRAM. Further, the interface to the host is also implemented in the logic layer. The commands sent from the host processor are simplified compared to traditional DRAM, leaving the management to the high performance logic layer.

The data communication between the controller and the DRAM arrays are managed in vertical slices. The cube of stacked memory and logic layers is partitioned vertically into slices, where each slice consists of rectangular die areas, one from each layer. One such vertical slice is also called a vault. The vault memory controller in the logic layer is responsible for accessing the data within the same vault. A global memory controller coordinates the vault controllers to access data in different vaults, with communication on the logic layer. The interface with the host processor, as well as with other 3D memory chips, is also implemented in the logic layer. The communication protocol between the host and the 3D DRAM is simplified compared to traditional DRAM, as the DRAM timing specifications are managed within the logic layer. The interconnect between the 3D DRAM chips can have different topology depending on usage scenarios [26]. A daisy chain with host processor is suitable when the number of DRAM chips is small; otherwise, the DRAM chips are connected in 2D mesh, and the host processor interfaces with a gateway DRAM chip.

Two major examples of 3D stacked memory are Hybrid Memory Cube (HMC) [27] and High-Bandwidth Memory (HBM) [28]. They were developed and standardized by different entities, but they share a similar design as described above. Figure 3.2 shows the architecture of HMC, with the 3D-stacked structure of layers and the various control modules in the logic layer. One major difference between HMC and HBM is the interface to the host processors. HMC

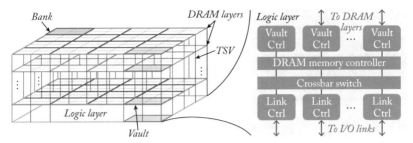

Figure 3.2: 3D stacked DRAM: HMC architecture [29].

uses packet-based serial links, which is more suitable for CPU hosts for ease of programming and control. On the contrary, HBM communicates with the host processor through silicon interposers with parallel links, and thus is more widely used for GPU processors which have higher degree of parallelism.

The 3D stacked DRAM solves the challenges met by earlier NMP designs in integrating the logic and DRAM process, as the DRAM and logic are in separate dies now. Many near-DRAM processing designs are thus inspired, targeting data-intensive application domains such as data analytics, graph processing, and deep learning.

3.2.2   NEAR-MEMORY PROCESSING ON 3D DRAM

The 3D DRAM provides an efficient base for near memory processing. The logic layer is implemented with high-speed logic process, so any NMP logic can be instantiated here. At the same time, the logic layer may communicate with the DRAM modules with high data transmission rate in TSV, thanks to the 3D-stacked structure [30].

Figure 3.3 shows a canonical architecture of near-memory processing with 3D-stacked DRAM. The processing elements (PEs) are usually integrated in the logic layer of the cube. The PE structure is application specific. Many PEs consist of arithmetic units and data buffers in SRAM [31, 32]. The PEs access the memory data through the memory interface which is also on the logic layer.

As the entire cube is partitioned into vaults, identical PEs are placed in every vault. Each PE can access the memory data in the same vault directly with the TSV. There is an interconnect for all the PEs in the logic layer for exchanging data. The PE can also access the memory data in other vaults through the interconnect. A Network-on-Chip (NoC) with mesh topology is usually chosen for the interconnect [33], as this matches the distribution of the vaults.

We now discuss a few variations of the above canonical NMP architecture with 3D DRAM. Broadly, the variations can be categorized based on the amount of computation performed on the host, proximity of the host to the 3D DRAM and application specific customizations applied to the processing elements in the logic layer.

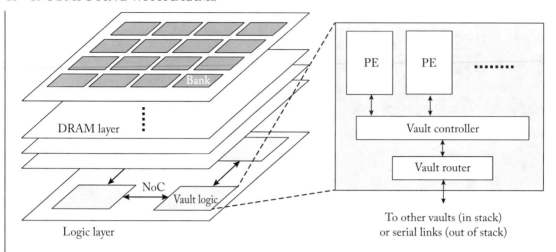

Figure 3.3: Canonical NMP architecture with 3D DRAM.

Variations Based on Interface with Host: In such a near-memory processing architecture with 3D DRAM, as the compute-capable 3D DRAM is usually treated as an accelerator or co-processor [34], the host processor is responsible for sending the near-memory processing instructions. Further, in some applications, the host processor also participates in the computation, on the less data-intensive part with high compute complexity. The typical accelerated applications include big data analytics [35, 36], graph processing [37], data reorganization [29], and neural networks [38–40].

On the other hand, the 3D DRAM can also have a tighter integration with host processor, instead of acting as a separated memory device [41]. For example, the 3D DRAM can be included as part of the memory system, and the host may decide whether to run an instruction as a conventional memory operation or using near-memory processing. Such a system needs additional support for cache coherence [42], instruction set extensions [41], and concurrent access from the host processor and the PEs [43].

Variations Based on PE Architecture: The PE can be in multiple formats apart from application-specific accelerators. For example, a group of general purpose CPU and GPU cores can be put in PEs [44], leading to ease of programming and reduced design complexity. PEs may consist of reconfigurable arrays of computing units [45], to create flexible near-memory accelerators. The PEs may apply various performance optimizations. For example, hardware prefetchers can be integrated into the PEs in order to better utilize the memory bandwidth [37].

Advantages and Shortcomings: In near-memory processing with 3D DRAM, the PEs are implemented in the high-performance logic process nodes, so there is no performance degradation. The overall performance also benefits from the improved latency and memory bandwidth of the 3D DRAM. Like all near-memory processing designs, the data movement between memory and processor is effectively reduced, leading to energy savings. Such a design also has room for improvements. The number of accesses to the DRAM modules is still not reduced although the data are processed close to the DRAM. Also, when an operation involves data from far away locations (in different vaults), the data still incurs remote transfer overheads before being processed.

3.3   COMMERCIAL NEAR-DRAM COMPUTING CHIPS

Recently, UPMEM realizes the idea of near-DRAM computation on a production chip [14]. The computation units are implemented in the DRAM process near the memory arrays. Each 64 MB DRAM array is paired with one processing unit, which runs at a 500 MHz frequency with a 14 stage pipeline, and has 24 hardware threads. Each DRAM chip contains one controller to orchestrate data transfer between the DRAM array and the processing unit via DMA operations. There are SRAM buffers for instructions and working data. The computing units support an ISA similar to RISC, including integer arithmetic, logical operations, and branch instructions. A special compiler maps C code to offload part of the computation with the near-DRAM computing units. Overall, the DRAM chip achieves a high memory bandwidth with near-DRAM computing, which can be used to accelerate a wide range of applications, such as DNA sequencing, data analytics, sparse matrix multiplication, and graph search.

The HBM-PIM [15] is a recent industry prototype work on near-memory processing based on HBM2 DRAM. Their main idea is to integrate programmable computing units into the DRAM dies near the memory banks. Such computing units enable the near-memory computation on the DRAM data, making data transfer to logic dies unnecessary. The computing units are placed adjacent to the memory cell arrays, and two memory banks share one computing unit. The computing units at multiple banks can work simultaneously to increase overall throughput. Each computing unit consists of instruction decoders, arithmetic units, scalar and vector register files, and the interface to handle conventional DRAM and specific control signals. The arithmetic unit supports floating-point addition and multiplication on a vector of FP16 data. The vector register files stores 256-bit entries as operands, while the scalar register file is used for control information. In HBM-PIM, the conventional memory controller is unmodified and a dedicated computing controller is designed to manage the computing units. To initiate near-memory computing, the host sends a row activate command with a specific row address. Then the sequence of instructions are sent to the computing unit for execution. Upon completion, the results of computing units are written to the memory cell arrays from the vector registers. The major advantage of the HBM-PIM is the combination of the high memory bandwidth with external logic provided by HBM, and the high internal computation throughput provided by the processing units in the DRAM dies. Such an architecture can be used for applications which require a large memory footprint and expose a high level of parallelism, such as machine learning models with a huge number of parameters. For example, the HBM-PIM achieves significant performance improvements and energy reduction for speech recognition benchmarks.

Pitfall Near-memory computing can significantly minimize the data movement cost.

This is not entirely true because of the following reasons.

- A large amount of energy is consumed by interconnects, which transfer data internally from a memory subarray to I/O. For example, DRAM's row activation takes 0.11 pJ/bit (909 pJ/row) for HBM2, while data movement and I/O take 3.48 pJ/bit at 50% toggle rate [46]. The data movement cost consists of pre-GSA (Global Sense Amplifier) data movement cost (43%), post-GSA data movement cost (34%), and I/O cost (23%). Neither in- nor near-memory computing can avoid the row activation cost and pre-GSA data movement cost. While in-memory computing can save the majority of post-GSA data movement cost and I/O cost, near-memory computing can only reduce the I/O cost unless it is placed very close to the memory array.

- For SRAM caches, the energy consumption of data movement is 1985 pJ while cache access consumes 467 pJ (L3 slice) [10]. Thus, H-Tree, which is the interconnect used for data transfer within a cache slice, consumes nearly 80% of cache energy spent in reading from a 2 MB L3 cache slice. In-memory computing can reduce the majority of the data transfer cost, while near-memory computing near an SRAM slice cannot.

- The savings of near-memory computing vary based on the proximity of the near-memory computing logic to the memory arrays. While a tight coupling of the logic and memory array can provide maximum cost savings for data movement to near-memory computing as well, it may not be optimal and cost-effective from other aspects such as memory density and process technology.

- Near-memory computing can reduce the latency to fetch data from memory. However, reduced communication overhead is often traded off by reduced computation throughput due to less performant cores or limited area for custom logic with enough parallelism and throughput.

3.4   IN-DRAM COMPUTING AND CHARGE SHARING TECHNIQUES

While there is a large body of work that explores near-memory computing using DRAM, including 3D stacked memory (previous section) and bit-serial ALUs attached to each bitline or sense amplifier [47, 48], it does not modify DRAM arrays for computation. There are several known obstacles for DRAM-based *in-memory* computing. For example, it is not always practical to have an ample amount of logic circuit within DRAM substrate because it could result in lower density. The computation throughput could also be limited by the datapath width. This section explores charge sharing techniques, one of the key enablers of DRAM-based in-memory computing. Charge sharing techniques activate more than one wordline and perform bitwise operations by exploiting altered charges in capacitors connected to the same bitline. Hence, it can provide some important logic operations with a small area cost. We present a typical operation sequence of DRAM reads, followed by the principle of triple row activation (TRA) of Ambit [49, 50], a representative in-DRAM computing work that introduces the charge-sharing technique. We then discuss some related work that enhances the charge-sharing techniques for lower cost and/or more supported operations.

3.4.1   DRAM READ SEQUENCE AND TRIPLE ROW ACTIVATION

We first present how the normal read operation of DRAM is performed. As shown in Figure 3.4, (1) both `bitline` and $\overline{\texttt{bitline}}$ are precharged to $1/2V_{DD}$, then (2) ACTIVATE command raises a wordline. If the capacitor is fully charged, the charge flows from the capacitor to `bitline`, and (3) `bitline` observes a voltage level with a positive deviation δ, i.e., $1/2V_{DD} + \delta$. Otherwise, it observes the opposite current flow and bitline voltage changed with a negative δ. Following that, (4) the sense amplifier senses the difference in the voltage level of `bitline` and $\overline{\texttt{bitline}}$, and amplifies the deviation to the stable state (i.e., `bitline` = V_{DD}, $\overline{\texttt{bitline}}$ = 0). At the same time, since the capacitor is still connected, (5) the sense amplifier drives the `bitline` and fully charges the capacitor if the deviation is positive.

Ambit [49, 50] proposes charge sharing-based bitwise AND and OR operation. Ambit simultaneously activates three wordlines (referred to as triple-row activation or TRA), as illustrated in Figure 3.5. Based on the charge sharing principles [51], the bitline deviation δ when

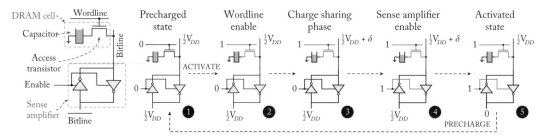

Figure 3.4: DRAM cell architecture and state transitions involved in DRAM cell activation [49].

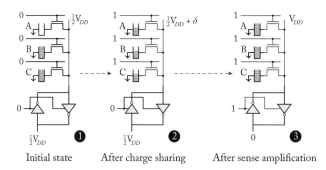

Figure 3.5: Triple-row activation and charge sharing [49].

multiple rows are activated is calculated as

$$\delta = \frac{kC_c V_{DD} + C_b 1/2V_{DD}}{3C_c + C_b} - \frac{1}{2}V_{DD} \tag{3.1}$$

$$= \frac{(2k-3)C_c}{6C_c + 2C_b} V_{DD}, \tag{3.2}$$

where C_c is the cell capacitance, C_b is the bitline capacitance, and k is the number of cells in the fully charged state. We assume an ideal capacitor (no capacitance variation, fully refreshed), transistor, and bitline behaviour (no resistance). According to Equation (3.2), the bitline deviation is positive (sensed as 1) if $k = 2, 3$ and negative (sensed as 0) if $k = 0, 1$. Therefore, if there are at least two fully charged cells before the charge sharing, V_{DD} is sensed, and since the sense amplifier drives the bitline to V_{DD}, all three cells will be fully charged. Otherwise, they will be discharged to 0.

The behavior of TRA is the same as a 3-input majority gate. Given A, B, and C represent the logical value of the three cells, it calculates $AB + BC + CA$, which can be transformed into $C(A + B) + \overline{C}(AB)$. Hence, by controlling C, TRA can perform AND ($C = 0$) and OR ($C = 1$). Ambit also supports NOT operation in a cell with an additional transistor, which enables the capacitor to be connected to $\overline{\texttt{bitline}}$ as in Figure 3.6. A combination of AND and NOT forms

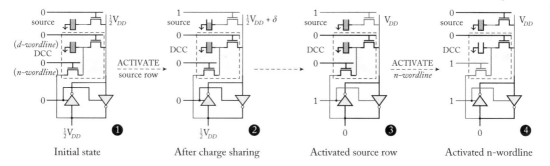

Figure 3.6: Bitwise NOT using a dual-contact cell [49].

NAND, a functionally complete operator. Therefore, Ambit can support any logical operations. Figure 3.7 shows the command sequences required to perform each Boolean operation.

3.4.2   ADDRESSING THE CHALLENGES OF CHARGE SHARING TECHNIQUES

The TRA-based charge sharing technique enables bulk in-DRAM logic operations with a small modification to the DRAM structure. However, it has the following challenges.

- **Customized row decoder and command sequence**: The peripheral circuit and memory controller interface need to be customized to perform TRA.

- **Limited operations**: The types of operations supported just by TRA are limited to AND and OR. NOT and other operations require an additional circuit. It also takes multiple TRA operations to construct non-native logic operations such as XOR, resulting in a large latency.

- **Sensitivity to charge variance**: Initial charges remaining in the cells and discharges during the operation can jeopardize TRA's accuracy.

- **Destructive operation**: After charge sharing, cell contents are overwritten. The copy operation is required to retain the original data.

In this section, we present some works that address these challenges.

Supporting Charge Sharing Without Circuit Modification   ComputeDRAM [52] demonstrates that off-the-shelf unmodified commercial DRAMs can perform charge-sharing-based computation. They manage to activate more than one wordline of a DRAM sub-array by manipulating command sequences violating the nominal timing specification and by activating multiple rows in rapid succession, as shown in Figure 3.8. They find multiple ACTIVATE commands interposed by PRECHARGE command can activate multiple rows within a timeframe

(a) Dk = Di **and** Dj

```
AAP (Di,  B0) ;T0 = Di
AAP (Dj,  B1) ;T1 = Dj
AAP (C0,  B2) ;T2 = 0
AAP (B12, Dk) ;Dk = T0 & T1
```

(b) Dk = Di **nand** Dj

```
AAP (Di,  B0) ;T0 = Di
AAP (Dj,  B1) ;T1 = Dj
AAP (C0,  B2) ;T2 = 0
AAP (B12, B5) ;DCC0 = !(T0 & T1)
AAP (B4,  Dk) ;Dk = DCC0
```

(c) Dk = Di **xor** Dj
= (Di & !Dj)|(!Di & Dj)

```
AAP (Di,  B8)  ;DCC0 = !Di, T0 = Di
AAP (Dj,  B9)  ;DCC1 = !Dj, T1 = Dj
AAP (C0,  B10) ;T2 = T3 = 0
 AP (B14)      ;T1 = DCC0 & T1
 AP (B15)      ;T0 = DCC1 & T0
AAP (C1,  B2)  ;T2 = 1
AAP (B12, Dk)  ;Dk = T0 | T1
```

or/nor/xnor can be implemented
by appropriately modifying the
control rows of and/nand/xor.

Addr.	Wordline	Addr.	Wordline	Addr.	Wordline(s)	Addr.	Wordline(s)
B0	T0	B4	DCC0	B8	$\overline{DCC0}$,T0	B12	T0,T1,T2
B1	T1	B5	$\overline{DCC0}$	B9	$\overline{DCC1}$,T1	B13	T1,T2,T3
B2	T2	B6	DCC1	B10	T2,T3	B14	DCC0,T1,T2
B3	T3	B7	$\overline{DCC1}$	B11	T0,T3	B15	DCC1,T0,T3

Mapping of B-group addresses to corresponding activated wordlines

Figure 3.7: Command sequences for bitwise Boolean operations and row address grouping [49]. The B-group addresses map to one of four designated rows T0–T3, d-/n-wordlines of two dual-contact cells (DCC0/1, $\overline{DCC0/1}$), or their combinations (trigger simultaneous activation of the wordlines), as listed in the table. The C-group addresses point to two pre-initialized rows: C0 (row with all zeros) and C1 (row with all ones). The D-group addresses corresponds to the rows that store regular data. Following command primitives are used: AAP = ACTIVATE–ACTIVATE–PRECHARGE, AP = ACTIVATE–PRECHARGE.

that the charge sharing is possible. Figure 3.9 shows their scheme to perform AND/OR operations. While their scheme does not support the NOT operation, they address it by storing negated data with the original data. Not all DRAM chips support charge-sharing by their manipulated command sequences, and not all columns always result in the desired computation result. However, their findings cast a new light that charge-sharing-based in-DRAM computation can be supported stably with minimal or no hardware changes to the DRAM DIMM.

Supporting More Operations   DRISA [53] proposes a 1T1C-based IM-P computing and 3T1C-based IM-A computing architecture (Figure 3.10). DRISA-1T1C uses the Ambit's charge sharing technique [50] to calculate AND and OR, and supplements them by adding latches and logic gates to the array peripherals. DRISA-3T1C changes the standard DRAM

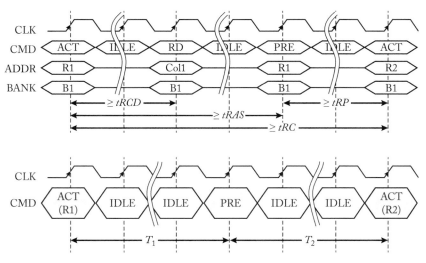

Figure 3.8: Typical command sequence for read (top) and ComputeDRAM's multi-row activation command sequence [52].

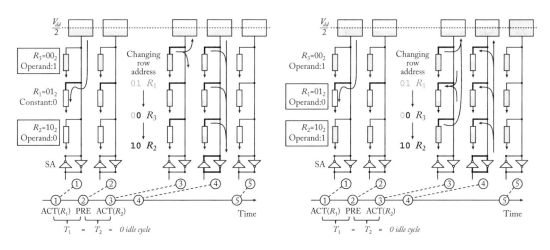

Figure 3.9: ComputeDRAM AND/OR operations [52].

cells to 3T1C and performs computation in the cells without adding other logic. The cell includes two separate read/write access transistors and an extra transistor that decouples the capacitor from the bitline. This extra transistor connects the cells in a NOR style on the bitline, providing the ability to naturally perform the NOR operation. DRISA also proposes adding a shift circuit to the array peripherals to support bit-parallel addition using a carry-save adder.

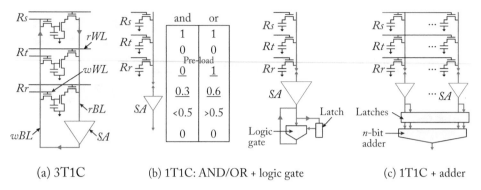

(a) 3T1C   (b) 1T1C: AND/OR + logic gate   (c) 1T1C + adder

Figure 3.10: DRISA architecture [53].

Addressing Charge Variance and Latency   While the charge-sharing-based techniques enable in-DRAM computing with minimum hardware modification, one of the downsides of them is that the initial charges remaining in the cells can jeopardize TRA's accuracy [54, 55]. To improve accuracy, Ambit copies two operands to the bottom rows near the sense amplifiers to keep the fully charged state. This copy operation is also essential to protect the original data from the destructive charge-sharing-based computation. Ambit also needs to write 1/0 to the third cell to define OR/AND operation before TRA. Although Ambit reserves several designated rows near the sense amplifiers for computation to reduce the number of copy operations, it still requires non-trivial cycles to deal with a complex Boolean operation (e.g., seven cycles to carry out XOR and XNOR, four cycles for AND).

ROC [56] takes a different approach of charge sharing to further reduce latency for logic operations, leveraging the characteristics of a diode connected with capacitors. As shown in Figure 3.11a, regardless of the initial state of cell C, it is charged to 1 if the transistor is connected to V_{DD}. Likewise, D will be discharged to 0 if the transistor is connected to ground. Hence, the current flow of C and D is unidirectional, allowing cells C and D to behave like OR and AND gates, respectively. For example, to perform $X + Y$, the content of X is copied to C, then Y is copied to D. If Y is 0, C's value remains the value of X; otherwise, C is set to 1. To perform an AND, first X is copied to D, then Y is copied to C. D will be 1 only if both X and Y are 1. ROC has a smaller latency than Ambit since it takes only two copy operations to compute the result. It also requires lesser area because ROC needs only two cells during computation. To make a functionally complete operator, ROC attaches one access transistor at the bottom of a column (Figure 3.11b), similar to Ambit. They also propose an enhanced ROC design with propagation and shift support by adding additional transistors and horizontal connectivity to the compute capacitors. By performing copy and computation simultaneously, ROC can reduce the computation cycles (e.g., four cycles for XOR, two cycles for AND), avoiding data corruption and result instability due to initial charges remained in cells.

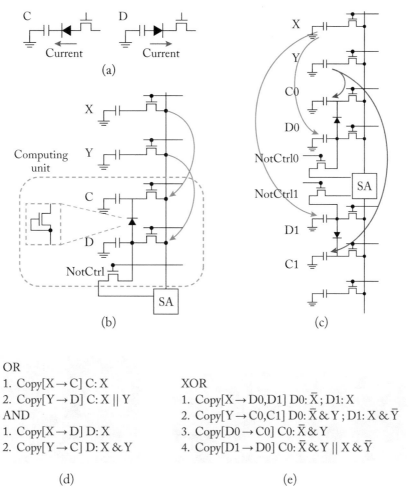

OR
1. Copy[X→C] C: X
2. Copy[Y→D] C: X || Y
AND
1. Copy[X→D] D: X
2. Copy[Y→C] D: X & Y

(d)

XOR
1. Copy[X→D0,D1] D0: $\bar{X}$; D1: X
2. Copy[Y→C0,C1] D0: $\bar{X}$ & Y ; D1: X & $\bar{Y}$
3. Copy[D0→C0] C0: $\bar{X}$ & Y
4. Copy[D1→D0] C0: $\bar{X}$ & Y || X & $\bar{Y}$

(e)

Figure 3.11: ROC architecture and command sequences [56].

3.4.3   DISCUSSION

The market for conventional DRAM is highly mature, and the next generation will likely see the prevalence of high bandwidth stacked memories and the emergence of compute capable DRAM. To preserve the density, speed, and stability as the core competencies of DRAM technology, potential in-memory architecture candidates should not pose enormous challenges for redesigning and re-validating DRAM technology which has matured over decades of production. We speculate that the drastic change in cell design (e.g., IM-A architecture) may not be practical from this point of view. There is also a challenge to make sure in-memory approaches work across process corners and temperatures. Typically, DRAM has no margin left after considering corners and temperatures; hence, in-memory methods that work in typical conditions would not work in practice, requiring some trade-offs to be made (e.g., increasing capacitor size or reducing refresh time). Still, recent research has found just a minor modification in the peripheral and/or command sequence can transform DRAM into an in-memory computing device. With compelling use cases of these operations in critical workloads and the development of a programming model, we envision these in-memory techniques should gather more attention in the market of commercial DRAMs.

CHAPTER 4

Computing with SRAMs

In the previous chapter we saw how the DRAM can be used for in-/near-memory computing. The Static RAM (SRAM) has a shorter data access latency compared to DRAM, and is also a mature memory technology. In this chapter we will explore the opportunity and challenges for in-SRAM computing. We will begin with the basics for SRAM in Section 4.1, followed by the digital-based in-SRAM computing approaches in Section 4.2, and analog/mixed-signal-based approaches in Section 4.3. Section 4.4 briefly discusses near-SRAM computing.

4.1   SRAM BASICS AND CACHE GEOMETRY

The SRAM is one important type of memory widely used in today's computer systems. The data stored is static, and periodic refreshing is not needed. The data in SRAM is also volatile so SRAM is a type of temporary data storage. SRAMs are used for the on-chip cache and register files of most modern CPUs. SRAM has larger and more complex bit cells compared to DRAM. Therefore, SRAM has lower data density and higher manufacturing cost, and is usually used for memory structures with smaller capacity. On the other hand, SRAM access speed is faster than DRAM. So SRAM is suitable for the high-speed cache and registers that are in the top part of the memory hierarchy.

Typical SRAM has a 6-transistor cell structure shown in Figure 4.1. The cell is made of a pair of cross-coupled inverters and two access transistors. The cross-coupled inverters have two stable states and thus can store one bit of data. The access transistors are used for reading and writing of the data bit.

The SRAM cells are arranged in an array for efficient data access. In an SRAM array, one wordline spans one row and connects to all cells in the row via the access transistors. Likewise, one pair of bitlines (BL/BLB) spans one column and connects to all cells in the column. Some peripheral logic is placed around each SRAM array, assisting the data access from the array. A row address decoder is connected to all wordlines, and is used to activate the correct wordline according to the row address. The BL peripherals include sense-amplifiers, bitline prechargers, and write drivers. In a read cycle, the bitlines are first precharged as a preparation for reading. Then the target wordline is activated, and at each bitline pair, either BL or BLB keeps high, depending on the corresponding bit-cell value. In a write cycle, each write driver raises either BL or BLB to the high voltage depending on the desired value to write. Then the target wordline is activated and the cross-coupled inverters at each cell change to the new state according to the BL/BLB value.

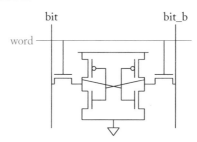

Figure 4.1: Structure of a 6T SRAM cell [57].

The SRAM cells must keep the stored data correct during the reading and writing process, as well as in the data holding stage. A quantitative measure of the stability of the SRAM cell is the noise margin, which is the maximum level of electrical noise that does not cause the data corruption in the SRAM cells.

We provide a brief overview of a cache's geometry in a modern processor. Figure 4.2 illustrates a multi-core processor modeled loosely after Intel's Xeon processors [58, 59]. Shared Last Level Cache (LLC) is distributed into many slices (8-14 for Xeon E5 discussed here), which are accessible to the cores through a shared ring interconnect (not shown in figure). Figure 4.2b shows a slice of LLC cache. The slice has 80 32-KB banks organized into 20 ways. Each bank is connected by two 16 KB sub-arrays. Figure 4.2c shows the internal organization of one 16 KB sub-array, composed of 8 KB SRAM arrays. Figure 4.2d shows one 8 KB SRAM array. A SRAM array is organized into multiple rows of data-storing bit-cells. Bit-cells in the same row share one word line, whereas bit-cells in the same column share one pair of bit lines.

In-SRAM vector arithmetic operations within the SRAM arrays (Figure 4.2d) can exploit the massive parallelism available in the cache structure by re-purposing thousands of SRAM arrays (4480 arrays in Xeon E5) into vector computational units. We observe that LLC access latency is dominated by wire-delays inside a cache slice, accessing upper-level cache control structures, and network-on-chip. Thus, while a typical LLC access can take ~30 cycles, an SRAM array access is only 1 cycle (at 4 GHz clock [58]). Fortunately, in-SRAM architectures require only SRAM array accesses and do not incur the overheads of a traditional cache access. Thus, vast amounts of energy and time spent on wires and higher-levels of memory hierarchy can be saved.

4.2 DIGITAL COMPUTING APPROACHES

Section 4.1 introduced the structure of the LLC and the opportunity of repurposing the SRAM arrays as computational units. LLC-based in-SRAM computation has its unique advantages over other computing platforms such as dedicated accelerator chips. First, the data movement between the CPU and the computational units is reduced. For a system with a dedicated co-processor such as a GPU, there is a communication overhead as data are transferred between the

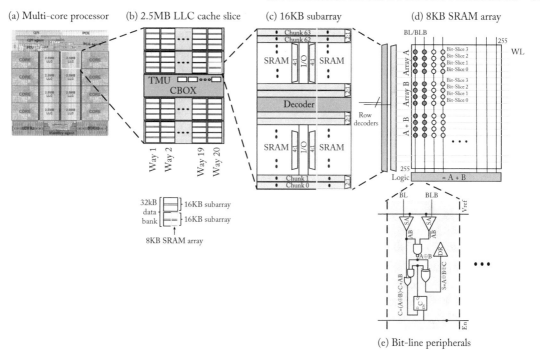

Figure 4.2: Cache geometry overview (adopted from [11]). (a) Multi-core processor with 8-24 Last Level Cache (LLC) slices. (b) Cache geometry of a single 2.5 MB LLC cache slice with 80 32-KB banks. Each 32-KB bank has two 16-KB sub-arrays. (c) One 16-KB sub-array composed of two 8-KB SRAM arrays. (d) One 8-KB SRAM array re-purposed to store data and do bit line computation with transposed data mapping for vector A and B. (e) Peripherals of the SRAM array to support computation.

host memory and device through the PCIe bus. In-cache computing does not involve an additional device so the above bottleneck is avoided. Second, in-SRAM computing is cost efficient. Although a hypothetical tighter integration of the CPU and the co-processor may mitigate the above communication problem, the overall area can be too large (for example, the Titan Xp GPU is 471 mm^2 in 16 nm). In contrast, the additional area required for in-SRAM arithmetic is 15.8 mm^2 in 22 nm, while providing 9× more computing resources than the GPU above. The in-SRAM computing is also more power efficient in terms of the Thermal Design Power (TDP). The TDP for the Xeon processor including the computing units is 296 W, while that for GPU is 640 W. Third, in-cache computing enables a more efficient and flexible on-chip resource partition between computation and storage. Compared to the GPU with a small on-chip memory, the CPU cache can be flexibly partitioned and provide a large buffer space for the memory-bound applications. On the other hand, for compute-intensive workloads, a large number of computing units can also satisfy the requirements.

In the rest of this section, we will discuss the in-SRAM computing details, including the logical operations supported in SRAM, the bit-serial algorithms for integer/floating-point arithmetic, and the mechanisms for data transposition.

Logical Operations   The logical operations in SRAM perform element-wise logical operations (e.g., AND, OR, NOR) on the data stored on multiple wordlines in an SRAM array. In-SRAM computing can be implemented by activating the wordlines simultaneously and sensing on the pair of bitlines [60]. Element-wise AND and NOR are the two basic operations, as they are the direct results of sensing BL and BLB. Specifically, at a compute cycle, two wordlines are activated, and if one bit-cell stores the data "0", the precharged bitline (BL) will be pulled down to below V_{ref}. If all the bit-cells store the data "1", the BL will remain at the precharged high value. By sensing the voltage on BL with a single-ended sense-amp, the logical AND can be computed on the data stored in the bit-cells. Similarly, the logical NOR can be computed by sensing the bitline-bar (BLB), as only when all the data are "0", the complement data are all "1" and the BLB will remain high. The extra hardware required includes the additional row decoder to activate the wordlines, and the single-ended sense-amps to separately sense both the BL and BLB attached to a bitcell (as opposed to the differential sense-amps in a normal SRAM array). Figure 4.3 provides a summary of the implementation of in-SRAM logical AND and NOR operations. When compared to the charge sharing based in-DRAM computing techniques discussed in the previous chapter, the bitline computing technique describe above for SRAM is not destructive (i.e., does not require copying of operands prior to operation). Also, since single-ended sense amplifiers are used, two separate logic operations can be performed in BL and BLB simultaneously.

During the in-SRAM computation, as multiple wordlines are activated and thus the circuit conditions are not the same as the normal SRAM data access, it is necessary to evaluate the robustness of the computation. Consider the following scenario. Two bit-cells in the same bitline store data value "0" and "1", respectively. During the computation cycle, as the bitline discharges, the bit-cell with value "1" is in a condition similar to normal write, with the BL discharging and the access transistor on. As the discharging bitline is well above 0, the incurred disturbance is relatively low. Also, as only two rows are activated, data corruption is less likely than activation on more rows. Further, several modifications to the SRAM array can mitigate the disturbance. The wordline is under driven to weaken the access transistors, and the power lines supplying the cross-coupled inverters in the cells are kept high. Such a modified wordline voltage prevents data corruption during the in-SRAM computing process.

With the sensed elementwise AND and NOR results of two wordlines, logic can be added to the array peripherals to implement other logical operations such as NOT, OR, and XOR [10]. For example, A XOR B = (A AND B) NOR (A NOR B). More complex operations such as

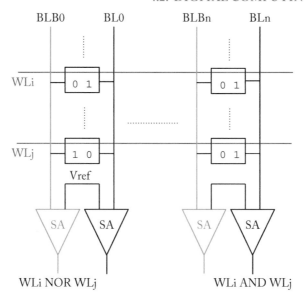

Figure 4.3: In-SRAM logical AND and NOR operations [10].

"compare" can also be implemented; wired-NORing all the bitwise XOR results produces the result for "compare."

Bit-Serial Arithmetic for Integers   The arithmetic operations in SRAM is built out of the logical operations discussed above.

A difference in the computation pattern is that arithmetic computation is done in the bit-serial format, as opposed to the bit-parallel format where the multiple bits of an operand are read out and processed as a whole. For bit-serial computation, the data are mapped to the SRAM arrays in a transposed layout. In the transposed mapping, all the bits of a data word are mapped to the same bitline, instead of to the same wordline as in a normal mapping. The two operands and the operation result are all mapped to the same bitline with all their data bits. The bit-serial algorithm processes one pair of bits from the two operands every cycle, starting from the least significant bits. Additional logic gates and registers are integrated to the bitline peripherals, serving as additional processing elements for a step in the bit-serial algorithm, such as a 1-bit full adder. At each bitline, one result bit can be written back every cycle. As a result of data reading, processing, and writing back every cycle, the in-SRAM computation happens at a lower frequency than normal data read/write operations.

Such a bit-serial computing format is more efficient than bit-parallel for in-SRAM computation. In the bit-parallel format for arithmetic, the carry bit needs to be propagated across the bitlines, resulting in extra communication difficulties. Instead, in bit-serial computation, each

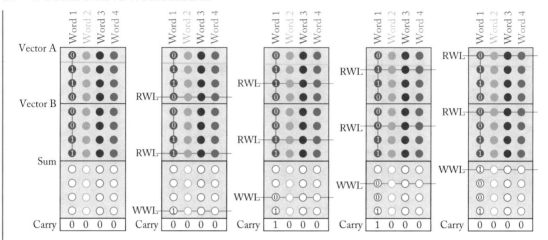

Figure 4.4: Bit-serial algorithm for in-SRAM integer addition [11].

carry bit is stored in a latch as part of the bitline peripheral, making the carry propagation fully parallel among all the bitline computing units.

There are other proposed in-SRAM computing techniques by analog computing (details in Section 4.3). As a comparison, the bit-serial computing in digital domain is CMOS compatible, and does not need the costly ADCs. Such a low overhead digital design is compatible and easier to adopt for the existing processors. Further, analog computing is usually for limited numerical precision, while bit-serial computing can flexibly support operands with reconfigurable precisions such as integer and floating point.

The above bit-serial processing diagram shares similarities with Associative Processing (AP), which is a historical idea and also has received recent attention for potential application in near/in-memory-computing [61, 62]. In AP, the computation is also performed in a bit-serial and word-parallel way. However, each single-bit operation is done by matching the pattern of the operands in a table and lookup for the results, while the in-SRAM computing implements the 1-bit operation with logic gates. Thus, traditional AP is based on CAM (Content-addressable memory), not SRAM.

Next, we discuss the detailed bit-serial algorithms for some typical arithmetic operations.

For bit-serial addition, one register is used for the carry bit. Every cycle, the full adder takes in one bit from each operand and the carry register, and calculates the sum bit and output carry bit. The carry bit register then updates its value, and the sum bit is written back to the array. Figure 4.4 shows the bit-serial addition algorithm, cycle by cycle, with 4-bit operands and 4 operations in parallel.

Bit-serial multiplication is implemented as a series of shifts and additions. One additional register (called tag register) is used to assist the calculation. The overall execution flow is a nested loop. In the outer loop, each bit of the multiplier is iterated. If that bit is 1, then the multiplicand is shifted and added to the product result. The if statement is implemented by predicated execution, using the tag bit as the condition. In the inner loop, first, the current bit of the multiplier is copied to the tag bit. Then the predicated accumulation of the multiplicand into the partial product result is done with bit-serial addition. The shift is performed implicitly by selecting the wordline index of the partial product result for accumulation. In such a bit-serial manner, it takes $n^2 + 3n - 2$ cycles to finish the n-bit multiplication.

One performance optimization of the bit-serial multiplication is leading zero search, which reduces the number of cycles for operands with a smaller effective bit width by skipping the computation on the leading zero bits. Further, the partial sum accumulation can be skipped for one iteration if all the bits from the multiplicand are zero.

Bit-Serial Arithmetic for Floating-Point Numbers   To support arithmetic for floating point, the bit-serial algorithm is modified and expanded from the integer arithmetic operations.

For floating point multiplication/division, the algorithm applies bit-serial addition/subtraction on the exponent bits, bit-serial multiplication/division on the mantissa bits, and logical XOR on the sign bit.

Floating point addition/subtraction involves shift with varying offsets and is discussed step-by-step in detail below (with addition as an example). The five steps are illustrated in Figure 4.5. (1) First, to prepare for the addition, the leading 1 of the mantissa bits is prepended and the mantissa bits are converted to 2's complement format according to the sign bit. (2) Then the differences between the exponents (ediff[i] in Figure 4.5) are calculated. For negative differences, the two operands are swapped and the differences are negated. This is to reduce the number of unique exponent differences by about half (they will be iterated in Step 4). (3) Then the shift and addition for mantissa bits is done for all distinct exponent differences sequentially. This is based on the observation that the dynamic range of exponents is small so there are few distinct exponent differences. Before the iteration, a search procedure is performed to find out all the distinct differences in an array. A leading zero search is first done to determine the upper limit of the difference to be searched. The search for a specific difference value is implemented in two cycles: in the first cycle, all wordlines corresponding to the bit value 1 in the target difference are activated and logical AND is sensed on the bitlines. In the second cycle, all wordlines corresponding to the bit value 0 in the target difference are activated; logical NOR is sensed and further AND-ed with the result from the first cycle as the final search hit vector result. (4) For all the distinct difference values, the mantissa bits of the second operand are shifted and added with the first operand. In bit-serial computation, the shift does not take extra cycles as it can be implemented by activating the wordlines for the correct bit positions. (5) Finally, the sum is converted back to the signed format. If there is an overflow in the sum, it is right-shifted and the exponent bit is increased.

Transpose of Data   The bit-serial arithmetic requires that the data should be in transposed format in the SRAM arrays. It can be achieved by either software manipulation or dedicated

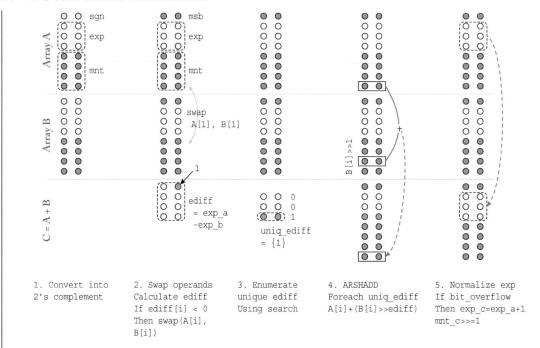

Figure 4.5: Bit-serial algorithm for in-SRAM floating point addition [63].

hardware units. Transposing in software requires programmer effort and can be used for the data operands that do not change at runtime.

Alternatively, the hardware Transpose Memory Units (TMU) can be applied. This option is desirable when the data are modified dynamically. The TMUs are placed in the cache control box (C-BOX in Figure 4.2b), and can read in data in the regular layout and send them out in transposed format (for the bit-serial computing operands), or vice versa (for the bit-serial results). The TMU can be implemented by an 8T SRAM array with two-dimensional access [64, 65]. The data access is supported in both vertical and horizontal directions with the extra access transistors in the memory cell. A few TMUs are sufficient to saturate the interconnect bandwidth among the SRAM arrays.

Pitfall In-SRAM computing reduces SRAM density and/or makes SRAM access time slow, so it is challenging to integrate in-SRAM computing into a CPU cache.

This is not true because of the following reasons.

- IM-A (in-memory (array)) computing that alters SRAM cell and array structure reduces SRAM density. However, many of the IM-P (in-memory (peripheral)) approaches keep SRAM arrays unmodified, and enable in-memory computation by adding a small amount of logic in the peripheral. Some IM-P approaches further claim they can be added to the original cache structure, leaving the cache functionality as-is.

- As stated above, many of the IM-P approaches (Section 4.2) keep the SRAM structure unmodified; thus, the cache access latency is not affected for regular reads and writes.

4.3   ANALOG AND DIGITAL MIXED-SIGNAL APPROACHES

The high availability of SRAM has attracted the attention of in-memory computing in SRAM bit-cells in the analog domain. These techniques are often referred to as mixed-signal computing as they perform analog computation in a bit-cell, while the computation result is converted into a digital value before it is referenced or written back to an SRAM array.

The SRAM in the conventional architecture and digital in-memory computing typically uses $L : 1$ column multiplexer (typically $L = 4 \sim 32$) to accommodate large sense amplifiers, as shown in Figure 4.9b. Thus, the number of bits fetched per single access to an array is limited to (N_{COL}/L) bits. On the other hand, mixed-signal computing is often composed of analog computation across bit-cells in different wordlines by multiple-row activation without involving sensing at the sense amplifiers (e.g., Figure 4.9a). Thus, a single access with α-row activation can process αN_{COL} bits, requiring much fewer precharge cycles to read out and process the same amount of data. While this brings energy and throughput gains, the analog nature of the mixed-signal bitline computation can reduce the fidelity/accuracy of data and computation [66].

Current-Based and Charge-Based Approaches   SRAM-based mixed-signal computing can roughly be classified into current-based approaches and charge-based approaches. The current-based approaches use the current-voltage (IV) characteristics of MOSFET in a bit-cell to perform multiplication. For example, Zhang et al. [67] store the operand (1-bit) being reused over iterations (e.g., weights in machine learning algorithms) in the SRAM array, and convert the other operand (5-bit) into a wordline voltage via DAC. The wordline voltage leads to corresponding bit-cell currents I_{BC}, as shown in Figure 4.6. Depending on the data stored in the bit-cell, I_{BC} is applied to either BL (bitline) or BLB (bitline bar), which can be thought of as multiplying I_{BC} with $-1/1$. The currents from all activated cells on a column are summed to-

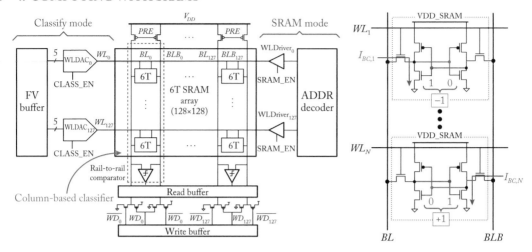

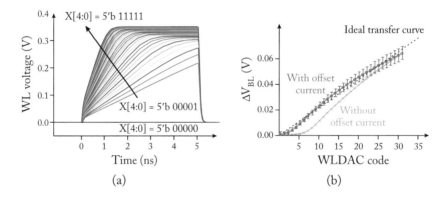

Figure 4.6: Multiplication and summation using a 6T SRAM bit-cell by a current-based approach [67].

Figure 4.7: (a) Transient WL pulses for different input codes and (b) bitline discharge transfer function vs. DAC input code [67].

gether in BL/BLB, resulting in bitline discharge and voltage drop. The amount of the voltage drop is proportional to the dot product of the input vector and the matrix data in the array. They also use a comparator to provide the sign of the dot product at the end of the computation. The current-based design is susceptible to substantial circuit non-idealities that stem from variations of bit-cells and non-linearities of the IV relationship. Figure 4.7 shows bitline discharge transfer function vs. DAC input code, illustrating the standard deviation (due to variations) and the nonlinearity.

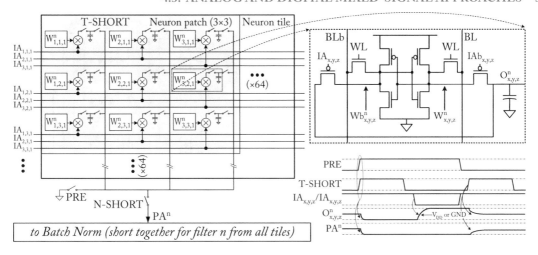

Figure 4.8: Charge sharing-based accumulation [68].

Charge-based approaches, on the other hand, use charge-sharing techniques to perform multiplication. For example, Valavi et al. [68] implements a capacitor connected to an 8T bit-cell to locally accumulate the product (Figure 4.8). Binary multiplication (XNOR or AND) is performed digitally on the 8T bit-cell, charging the local capacitor conditionally on the binary product. Then, the results from different cells are accumulated by charge sharing of local capacitors, controlled by the T-SHORT switches. Adding two transistors and a capacitor incurs $\sim 1.8\times$ area overhead compared to a standard 6T SRAM cell. While current-based design can be subject to a tight restriction on the neurons' precision due to a limited dynamic range of BL/BLB for I-V conversion, charge-domain accumulation can benefit from a larger dynamic range from the shorted capacitance. Moreover, capacitors are less susceptible to variations and non-linearity originating from temperature and process. Although they support only binary input, the accuracy loss of a fabricated chip compared to a software implementation is small (0.2–0.3%). Instead of using dedicated capacitors, some other designs use capacitance intrinsic to a bitline [69].

Improving Bit-Precision   One of the limitations of SRAM-based designs is bit precision. To support bit precision beyond a single bit, mapping a value to multiple bit cells is required because the SRAM bit-cell is essentially a single-level cell (SLC). Also, the input value has to be precisely modulated and fed to the rows with corresponding values. Ming et al. [66] adopt vertical data mapping and pulse width modulation (PWM) to support multi-bit operations. As shown in Figure 4.9a, data is stored vertically in a column, similar to the bit-serial computing approach discussed in Section 4.2. The input pulse is kept high for a time duration $T_i \propto 2^i$, which is exponential to the bit position of bit b_i. An example pulse sequence is illustrated in

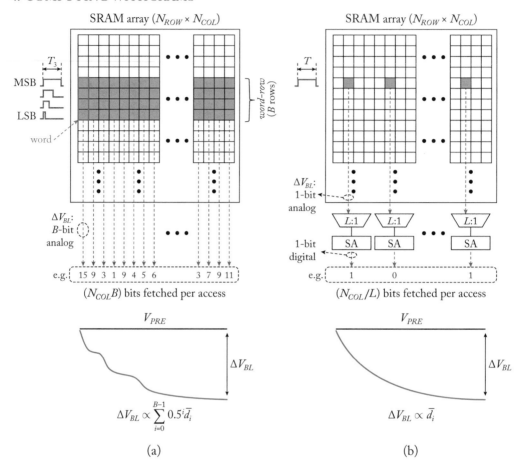

Figure 4.9: Multi-row activation in a mixed-signal architecture (a) and typical SRAM array architecture (b) (adoped from [66]).

Figure 4.10. This generates BL voltage drop ΔV_{BL} and BLB voltage drop ΔV_{BLB} as follows:

$$\Delta V_{BL}(D) = \frac{V_{PRE}}{R_{BL}C_{BL}}T_{\min}\sum_{i=0}^{n-1}2^i\overline{d_i}, \tag{4.1}$$

$$\Delta V_{BLB}(D) = \frac{V_{PRE}}{R_{BL}C_{BL}}T_{\min}\sum_{i=0}^{n-1}2^i d_i, \tag{4.2}$$

where $T_{\min}$, V_{PRE}, C_{BL}, and R_{BL} are the LSB pulse width $(= T_0)$, the BL precharge voltage, the bitline capacitance, and the nominal resistance of the BL discharge path. The input D is a n-bit decimal number and is given by $D = \sum_{i=0}^{n-1}2^i d_i$ (d_i is the i-th bit slice of D). For these

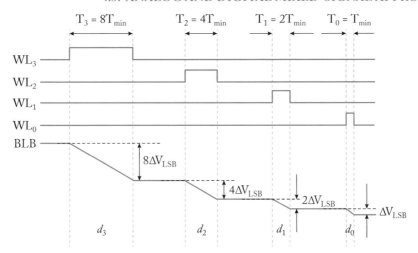

Figure 4.10: Input pulse sequence and bitline voltage drop of pulse width modulation [70].

equations to hold, the following conditions have to be met: (1) $T_i \ll R_i C_{BL}$ (R_i is the discharge path resistance of the i-th bitcell), (2) $T_i = 2^i T_{min}$, (3) $R_i = R_{BL}$ (no variation across rows), and (4) R_{BL} is a constant over V_{BL}. Without any modification to the SRAM array under these conditions, this enables computing sum of absolute difference [70], addition, and subtraction [66] of operands stored in the same column. An addition $A + B$ is naturally obtained by pulsing word-rows containing A and B giving

$$\Delta V_{BL} = \Delta V_{LSB} \sum_{i=0}^{n-1} 2^i (\overline{a_i} + \overline{b_i}) = \Delta V_{LSB}(\overline{A} + \overline{B}), \tag{4.3}$$

while subtraction requires B to be stored in a bit-flipped manner to use 1's complement. Alternatively, one can use a replica bit-cell [66] to facilitate access to flipped data.

Multiplication is performed in an external mixed-signal multiplication unit. Kang et al. [71] employ a redistribution based multiplier, which uses charge sharing of capacitors to perform division by two and accumulation and outputs a scaled product. Operand B is given in a binary format to the multiplier, and the sign bit is processed separately. A dot product is calculated by charge-sharing the multipliers' outputs in the participating columns using rails shared by the columns in the array. Positive and negative values are aggregated separately, and the sum of negative values is subtracted after the ADC.

Multi-row read using PWM is subject to several circuit non-idealities. One cause of non-linearity of the multi-row read process is from voltage-dependent resistance of the discharge path in access and pull-down transistors in a bit-cell. Also, the read precision can be affected by the local transistor's threshold voltage mismatch across bit-cells, which is caused by random dopant fluctuations [72] and finite transition times of PWM pulses. The analog multiplier can also have non-ideality. Considering these factors, Kang et al. [66] report that it is challenging to perform multi-row PWM activation with more than 4-bits. They propose a sub-ranged read technique that subdivides data with higher bit precision (e.g., $n = 8$-bit) into two parts, storing the upper-half word (BLM) and the lower-half word (BLL) in adjacent columns. This technique modifies the ratio of the bitline capacitances of BLM and BLL, $C_M : C_L$, to $2^{(n/2)} : 1$ by introducing a tuning capacitor. Charge sharing of BLM and BLL generates a bitline voltage drop

$$\Delta V_{BL}(D) = \frac{C_M \Delta V_{BLM} + C_L \Delta V_{BLL}(D_L)}{C_M + C_L} \tag{4.4}$$

$$= \frac{2^{n/2} \Delta V_{BLM}(D_M) + \Delta V_{BLL}(D_L)}{2^{n/2} + 1}. \tag{4.5}$$

Execution Model and ISA for Mixed-Precision Computing   PROMISE [73] extends the PWM modulated mixed-signal processing to support a variety of operations in machine learning algorithms. PROMISE has analog Scalar Distance (aSD) blocks that produce a scalar distance for each column using the mixed-signal processing, and analog Vector Distance (aVD) blocks that perform aggregation by charge-sharing all the analog outputs from the aSD blocks, as shown in Figure 4.11. PROMISE has a 4-stage pipeline: (1) aRead (analog read, aADD, aSUB); (2) aSD and aVD (compare, abs, square, mult); (3) ADC; and (4) TH (performs accumulation, mean, threshold, max, min, sigmoid, ReLU of digitized result from ADC). Pipelining analog value is supported by analog flip-flops that use a switch and a capacitor to temporarily hold the output. Instructions for each pipeline stage can be provided using a Very Large Instruction Width (VLIW) like instruction set architecture (ISA). PROMISE also proposes a compiler that takes programs written in Julia and outputs PROMISE ISA instructions, considering pipeline and delay constraints of the architecture. The compiler can explore accuracy vs. energy trade-offs by applying an analog error model, varying swing voltages (ΔV_{BL} in Figure 4.9) for aRead, and by encoding it in the ISA. The output of aRead follows a normal distribution $\hat{W} \sim \mathcal{N}(W, \sigma_W^2)$, and its standard deviation σ_W is inversely proportional to the swing voltage parameter.

4.3.1   DISCUSSION

One of the challenges in mixed-signal computing in SRAM is keeping adequate swing voltage (ΔV_{BL}) range to maintain both cell stability and accuracy. Too large swing voltages can cause read disturbance, which destroys cell content while reading. On the other hand, too small swing voltages can increase the chance that circuit non-ideality affects the computation results, thus reducing accuracy. While swing voltage can be tuned by wordline voltage, pulse width (for PWM modulated inputs), etc., the constraint of the swing voltage still limits the output resolution. Some works add another transistor to isolate the bit-cell from a large voltage swing. However, such designs lead to a larger bit-cell area and decrease bit-cell density.

   Cell density, throughput, and latency are important factors for efficiency, particularly for SRAM-based computing. SRAM inherently has large bit-cells compared to other memories,

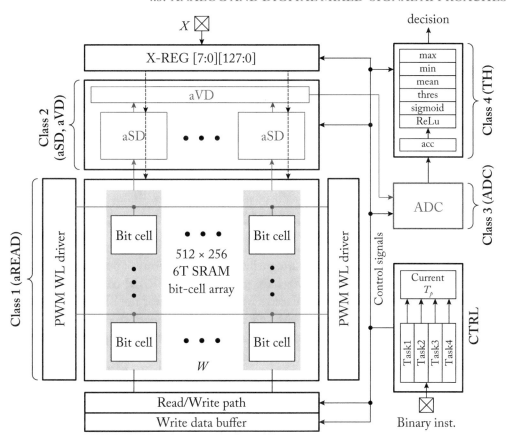

Figure 4.11: PROMISE architecture [73].

let alone SRAM-based IM-A architecture. Modifying the standard 6T SRAM cell for analog computation leads to low memory density, limiting the type of workload that can run on the device without external memory traffic. For example, among the architectures covered in this section, memory density (KB/mm², scaled to 65-nm process) is the highest for 6T-based design (30 for [67], 24 for [66]) and followed by the others (22 for 10T design [69], 16 for 8T1C design [68]).

Computation throughput is mainly dependent on density, precision, and accelerator design. Among these works, binary boost classifier [67] and binary CNN accelerator [68] (no ADC) achieved high throughput (536 and 1531 GOPS/mm², respectively), while the applicable workloads are restricted compared to others. Architectures employing 8-bit computation [66, 69] achieve 118 and 15 GOPS/mm². Therefore, throughput is also traded off by flexibility, i.e., to what extent the architecture is generalized to a broad category of workloads.

As for latency, mixed-signal computing on SRAM tends to achieve high clock frequency, especially when SRAM cells are unmodified (e.g., 1 GHz [66]). Thus, the latency (and energy) for a single arithmetic operation can be lower than the digital-based counterparts. However, a larger delay may occur for certain analog operations and ADC [73], both residing outside the array. These analog operations and circuitry can be a bottleneck in terms of latency, area, and throughput. For example, performing simple arithmetic and storing the results without aggregation may cause a severe bottleneck in ADC. To achieve more efficacy, reducing the number of A/D conversions is likely to be the key. The cost of analog circuitry also has to be carefully evaluated. Likewise, keeping density is equivalently important because external memory traffic can quickly extinguish the performance benefit, if frequent.

4.4   NEAR-SRAM COMPUTING

Near-SRAM computing places additional computing components near an organized SRAM structure, in contrast to in-SRAM computing where the SRAM array or peripheral are modified for computing. Earlier works in the 1990s have proposed near-SRAM computing devices as a co-processor. For example, Terasys [3] placed single-bit ALUs near the SRAM arrays, one per bitline. The data are read out from SRAM and computation is performed with the ALUs in a bit-serial way. The host processor is responsible for sending computing instructions to the SRAM arrays.

CPU cache is a major architectural component built with SRAM, and thus many recent works have explored the opportunities of near-SRAM computing with CPU cache. The computation is realized by the customized PEs placed in the CPU cache hierarchy. To achieve an adequate level of parallelism, PEs are placed near the last-level cache slices, on the data path between the core and the last-level cache slices [74]. The PEs have various types of implementation including simple in-order cores, reconfigurable fabric like FPGA [75], and customized logic for specific applications [76]. The applications accelerated are mostly data-intensive operations which involve CPU processing, such as database [76], and functions in the network stack [74]. There are optimizations for near-SRAM computing on simplified programming model and flexible computing scheduling. For example, Livia [75] optimizes the location of the computation (PE or CPU core) based on the locations of operands in the memory hierarchy, thus reducing the overall data movement.

Although the level of parallelism is lower than in-SRAM computing, near-SRAM computing does not need low-level modifications of SRAM structures, and supports more flexible computing patterns with different PE designs.

CHAPTER 5

Computing with Non-Volatile Memories

5.1 COMPUTING WITH MEMRISTORS

Emerging non-volatile memories (NVMs) have been an attractive memory substrate due to high density and the potential to replace DRAM main memory. Some advanced technologies of non-volatile memories use programmable resistive elements referred to as *memristors*.

Memristors are characterized by linear current-voltage (IV) characteristics called memristance. Memristance is defined in terms of the relationship between magnetic flux linkage Φ_m and the amount of charge that has flowed q, characterized by the following memristance function, which describes the charge-dependent rate of change of flux with charge:

$$M(q) = \frac{d\Phi_m}{dq}. \tag{5.1}$$

Using the time integral of voltage $V(t) = d\Phi_m/dt$ and the time integral of current $I(t) = dq/dt$, we obtain

$$M(q(t)) = \frac{d\Phi_m/dt}{dq/dt} = \frac{V(t)}{I(t)}, \tag{5.2}$$

$$V(t) = M(q(t))I(t). \tag{5.3}$$

By regarding memristance as charge-dependent resistance, we obtain a similar relationship as Ohm's law

$$V(t) = R(t)I(t), \tag{5.4}$$

and by solving for current as a function of time,

$$I(t) = V(t)/R(t) = V(t)G(t). \tag{5.5}$$

Various memristors use material systems that exhibit thermal or ionic resistive switching effects, such as phase-change chalcogenides and solid-state electrolytes. By applying a sufficiently high voltage, the memristor cell forms conductive filaments, which enables it to transition between high resistance (reset) state and low resistance (set) state. This internal state change is retained without power, providing non-volatility. The data is read by injecting the reference voltage and sensing the current from the memristor cell through the bitline, following the relationship of

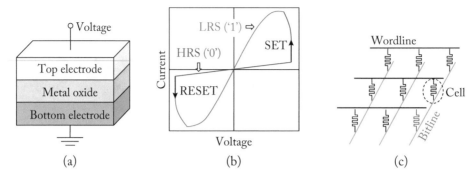

Figure 5.1: ReRAM cell and array architecture (adopted from [77]). (a) Conceptual view of a ReRAM cell; (b) I-V curve of bipolar switching; and (c) schematic view of a crossbar architecture.

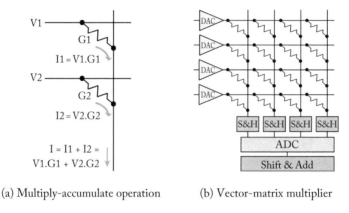

(a) Multiply-accumulate operation   (b) Vector-matrix multiplier

Figure 5.2: In-memory computing in ReRAM (adopted from [78]).

Equation (5.5). Due to the linear IV characteristics of memristor (Figure 5.1b), one cell can be programmed to 2^n different states (typically n is 1–5), and decode n-bit data by sensing the current magnitude. In other words, a memristor cell can function as a multi-level cell (MLC) device.

The linear IV characteristics of memristors are further exploited for in-memory computation in the analog domain (Figure 5.2). Changing the reference voltage within the region below the threshold voltages for set and reset still holds the memristance in Equation (5.3). As Equation (5.5) explains, the bitline current can be interpreted as the result of the multiplication of cell conductance and the input voltage. Furthermore, by activating multiple rows, currents that flow from different memristor cells sharing a bitline accumulate in the bitline, following Kirchhoff's law. This analog computing capability of memristors is first proposed for accelerating neural network workloads of which computation is dominated by multiply-accumulate (MAC) operations that compose dense matrix multiplications. For example, in such a system, weights are stored as the conductance of memristor cells, and a voltage proportional to the input activation is applied across the cells. The accumulation is performed in each bitline, as described above.

There are several favored NVM devices for in-memory computing: resistive RAM (ReRAM or RRAM), phase change memory (PCM), spin-transfer torque magnetic RAM (STT-MRAM), etc. In this section, we focus on ReRAM, the representative memory technology using memristors.

5.1.1   CHALLENGES IN USING MEMRISTORS

While there is a large body of work that exploits the compute capability of ReRAMs in recent years, this emerging memory technology is still speculative because of the unique challenges of memristor-based NVM devices. Below we list some of the important challenges in designing memristor-based in-memory accelerators.

- **Variation**: ReRAM-based devices are susceptible to device-to-device and cycle-to-cycle variations. They are caused by process variation (i.e., parameter deviation of devices from their nominal values), noise, non-zero wire resistance, nonlinearity in I-V characteristics, resistance drift (i.e., time-related deterioration of the resistance), etc. Especially a large ReRAM array with heavy wire congestion suffers from a high defect rate and poor reliability of read and write operations due to voltage-drop and process variation. The nonlineality of the I-V curve leads to non-linear conductance across the conductance range. This affects the number of bits stored per cell and the signaling method for input and output.

- **Operation cost**: The energy and latency of programming memristors can be much higher than those of volatile memories. ReRAM devices require a higher voltage to perform set and reset operations. Also, due to the variation, some ReRAM devices need write validation to verify the integrity of cell contents after each program operation. In particular, neural network training and error-intolerant workloads that need precise calibration of ReRAM cell contents require frequent write operations, leading to considerable power consumption.

- **Write endurance**: ReRAM devices typically have severely limited write endurance (cell lifetime) compared to the volatile memories. Aged cells are more susceptible to conductance nonlinearity and write errors.

- **ADC/DAC**: Analog computation in ReRAM necessitates the use of Digital-to-Analog Converters (DACs) and Analog-to-Digital Converters (ADCs) to communicate with digital circuitry. However, ADCs and DACs incur severe area/energy overheads and can degrade the signal precision. In particular, the area and power cost for

ADC exponentially grows with the ADC resolution (bits per analog signal), impacting the overall memory density and performance of ReRAM accelerators. ADC/DAC can take 85–98% of the total area and power of a neuromorphic computing device.

- **Limitation in data representations and operations**: Since ReRAM devices cannot store negative resistance, it is not straightforward to support negative weights in a neural network. Likewise, some neural network layers cannot be implemented in ReRAM crossbars (e.g., Local Response Normalization (LRN) layers).

In the following section, we will discuss various approaches proposed to tackle these challenges.

5.1.2   RERAM-BASED IN-MEMORY COMPUTING

Many ReRAM-based in-memory accelerator studies propose techniques for accelerating neural networks. The key kernel of widely used neural networks is composed of dot-product operations. The detailed workload description of the neural networks is described in Section 6.1. This section walks through fundamental approaches to enable in-memory arithmetic and their unique optimizations for ReRAM-based architectures.

Data Representation

ReRAM-based neural network computation has employed 1-bit to 12-bit synaptic weights [79–81]. While semiconductor research has been improving the precision of MLC ReRAM cells, it is still not enough to reliably support the required precision (e.g., 8-bit for image data) of popular neural networks. The Dot-Product Engine from HP Labs with 256×256 crossbar array, for example, achieves 6-bit output precision from 4-bit synaptic weights and 7-bit output precision from 6-bit synaptic weights [82], considering the effect of ReRAM noises. While the increased precision of MLC cells can potentially provide higher density and compute efficiency, challenges are still present because it increases the chance of accuracy getting affected by analog noises, process variation, and nonlinearity of cells. Likewise, it necessitates high-resolution ADCs/DACs.

One solution is to let several ReRAM cells of smaller precision compose a value. PRIME [77] uses two 4-bit ReRAM cells to represent 8-bit synaptic weights, and two 3-bit input voltages (fed from DAC) to represent 6-bit inputs. ADC decodes the higher 6-bits of the dot-product. A peripheral circuit then combines the 6-bit products of all four combinations of two high-bit parts and two low-bit parts using adders and shifters.

ISAAC [78] introduces several data representation techniques to support precise 16-bit dot-product operation on ReRAM arrays with a conservative MLC parameter. A 16-bit synaptic weight is stored in $16/v$ v-bit cells in a single row (e.g., $v = 2$), and a 16-bit input is sequentially fed from DAC using 16 binary voltage levels. Since DAC needs to read a single bit of the 16-bit number, 1-bit DAC suffices. The product currents are accumulated, ADC-converted, and digitally merged in a pipelined manner over the 16 sequential operations. Assuming a crossbar array of R rows, precisely accumulating R multiplications of v-bit inputs and w-bit synaptic weights necessitates ADC resolution A as follows:

$$A = \begin{cases} \log(R) + v + w, & \text{if } v > 1 \text{ and } w > 1 \\ \log(R) + v + w - 1, & \text{otherwise.} \end{cases} \tag{5.6}$$

To further reduce the size of the ADC, the w-bit synaptic weights W are stored in an inverted form $\overline{W} = 2^w - 1 - W$ if the weights in a column are collectively large (i.e., with maximal inputs, the sum-of-products yields the most significant bit (MSB) of 1). This encoding ensures that the MSB of the sum-of-products is always 0 and reduces the size of ADC by one bit. ADCs have the most significant contribution in the overall area and power, and there is a nearly exponential relationship between the resolution and the cost of ADC. Thus, this optimization has a non-trivial impact on the overall efficiency. ISAAC enables 128 element 16-bit dot-products in 18 cycles using 8-bit ADC (each cycle takes 100 ns).

Supporting negative weights is also an important problem to solve. Many works store positive and negative weights in different ReRAM arrays [77, 83, 84], or use only non-negative weights [85], because ReRAM cannot store a negative weight. ISAAC [78] uses 2's complement representation for inputs and a biased format x_{b16} (e.g., $-2^{15} = 0_{b16}, 0 = 2^{15}_{b16}, 2^{15} - 1 = (2^{16} - 1)_{b16}$) for synaptic weights. The bias of 2^{15} is subtracted from the end result for each 1 in the input.

Another research direction leverages neural networks whose data representation is suitable for ReRAM crossbars. Many circuit works adopt ternary state (1.5-bit) to eight state (3-bit) ReRAM cells due to the semiconductor-level challenges mentioned above. Likewise, digital (binary) ReRAM cells have been explored for in-memory computing to maximize reliability and memory density. These works run neural networks with a very small-sized weight (e.g., binary-based neural networks). Chen et al. [86] demonstrate a binary-input ternary-weighted (BITW) neural network on ReRAM arrays with 32×32 binary state cells. Negative weight (-1) is stored in a separate array macro. They propose a circuit level technique called distance-racing current-sense amplifier to overcome the large input offset issue and reduced sensing margin. Xue et al. [87] report a fabricated 55 nm 1 Mb computable ReRAM macro that runs networks of 2-bit inputs, 3-bit weights, and 4-bit outputs with a small computation latency (14.6 ns).

Supporting Various Arithmetic Operations
While many ReRAM based in-memory computing works leverage the analog multiply and accumulate (MAC) functionality of crossbars, this section presents works that propose various arithmetic primitives for supporting a variety of applications and enhanced precision.

Arithmetic and Logical Operations Pinatubo [88] presents a technique to perform bulk bitwise logic operations such as AND, OR, XOR, and NOT, in single-level cell (or binary)

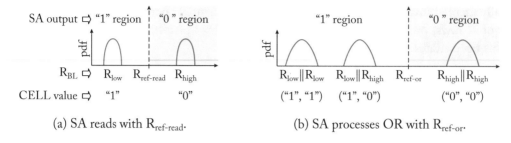

(a) SA reads with $R_{ref\text{-}read}$.   (b) SA processes OR with $R_{ref\text{-}or}$.

Figure 5.3: Modified reference values in SA to enable logic operations in Pinatubo [88].

ReRAM. While conventional read access activates a single wordline to compare the resistances against a reference resistance value to determine the state (0 or 1) of the stored value, their technique simultaneously activates two rows and reads the resistance of two cells forming a parallel connection. For computing AND and OR, only the reference voltage has to be changed as shown in Figure 5.3. For example, OR is calculated using a reference resistance R_{RefOR}, which is larger than the value for the normal read $R_{RefRead}$. XOR requires two micro-steps: one operand is read to a capacitor and the other is read to a latch, then the output of two add-on transistors provides the XOR result. Inversion (NOT) uses the differential value from the latch. It is also possible to extend OR operation to compute a reduction OR for multiple rows at a time. The output can be sent to the I/O bus or written back to another memory row. Pinatubo requires changes to the sense amplifiers and row driver, thus it is classified into an IM-P architecture. While applications need to align data for in-memory computation, it achieves higher energy efficiency and performance compared to conventional architectures and DRAM-based PIM techniques.

Various IM-A based architectures that support logical operations are presented later in the subsection *gate-mapping techniques*.

Many works support analog addition, subtraction, and multiplication on ReRAM, as mentioned in the previous sections. They use the shifter and adder unit in the peripheral circuit for carry propagation and partial result summation, taking advantage of the relatively large latency of the read access to the ReRAM crossbar. This may be a bottleneck with a ReRAM crossbar with faster read access, such as binary ReRAM. Imani et al. [89] propose performing addition and multiplication operations using a reconfigurable interconnect implemented within an array of ReRAM crossbars. This interconnect inherently supports shift operations, and by using a carry-save adder circuit implemented using MAGIC gates (see the gate mapping section), it performs addition and multiplication on the ReRAM crossbar array.

Floating Point Operations Some classes of applications, such as scientific applications, require high-precision floating point operations. On the other hand, ReRAM based in-memory computing is essentially a fixed-point computing device, thus supporting higher precision in

ReRAMs necessitates special support. Feinberg et al. [90] propose an architecture that converts floating point into fixed point in ReRAM crossbars, preserving precision. A double precision data is converted into a 118-bit fixed point number and error correcting code, then stored into 127 crossbars with each crossbar containing a single bit of the data. Addition and multiplication are performed in a similar way as ISAAC, except that a bit slice of the multiplicand is fed to all 127 crossbars in a cluster. The peripheral circuit implements a reduction tree of shifters and adders to aggregate the results and converts them into a floating point representation. To avoid extraneous calculations, they exploit the exponent range locality. This is based on the observation that a large dynamic range of exponents in the result is unlikely in many applications. Also, they do not perform all 127×127 crossbar operations for producing a product. Since every bit slice in the crossbars must be multiplied by every bit slice of the multiplicand fed from a driver, many of these operations would contribute only to the portion of the mantissa. Thus, once sufficient mantissa bits are calculated, it is safe to terminate the partial product summation upon observing the condition that no carry bits from lower bits will impact the mantissa bits.

FloatPIM [91] directly supports floating-point representations. To perform floating point addition, one needs to normalize the mantissa for the operand with a larger exponent value, if there is a difference in the operands' exponent. FloatPIM uses binary ReRAMs and performs bit-serial floating point operations in a similar way, as presented in Chapter 4.2. It also utilizes NOR operation based subtraction and exact search operations to calculate and identify the sign bit of the exponent difference.

Reducing the Overheads of Analog Computing

A/D converters are significant contributors to power and area. ISAAC [78] reports a single ADC takes about 48x larger area compared to a ReRAM array. To reduce the area for analog computation, several approaches have been proposed, and they can be categorized into three approaches: (1) avoid A/D conversion, (2) reduce the number of A/D conversions, and (3) use simpler A/D converters.

PipeLayer [92] avoids A/D conversion by using spike drivers and Integration and Fire circuit. For an N-bit input value, they use N time-slots to send a spike from a spike driver, which generates voltage ranging from $V_0/2^{N-1}$ to $V_0/2$ based on the bit position of the input. Integration and Fire circuit has a capacitor that accumulates current and emits spikes when it gets charged. By counting the number of spikes, they can identify the products of input data and weights in the crossbar.

CASCADE [93] reduces the number of A/D conversions by connecting compute ReRAM arrays with buffer ReRAM arrays to extend the processing in the analog domain. They use trans-impedance amplifiers (TIA) as the interface to convert analog currents of compute ReRAMs' bitline outputs into analog voltages that can be directly fed into buffer ReRAMs as inputs. TIA enables the partial sums from the compute ReRAM to be stored in the buffer

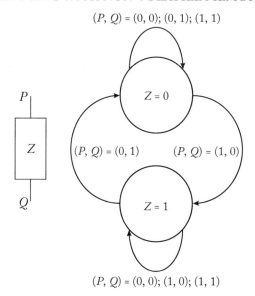

P	Q	Z	Z_n
0	0	0	0
0	1	0	0
1	0	0	1
1	1	0	0

$$Z_n = P \cdot \overline{Q}$$

P	Q	Z	Z_n
0	0	1	1
0	1	1	0
1	0	1	1
1	1	1	1

$$Z_n = P + \overline{Q}$$

Figure 5.4: Stateful material implication logic and majority gate function of a single memristor (adopted from [96]).

ReRAM, and the accumulated sum is calculated using a minimum number of A/D conversions (e.g., 10 A/D conversions for data with 6- to 10-bit resolution).

Gate Mapping Techniques

Gate mapping techniques map the functionality of binary logic gates on memristor cells and their connectivity. A single memristor can implement stateful material implication logic [94, 95], using its polarity. Assuming a memristor with binary states, its state transition will be determined based on the voltage (positive or negative) applied to the top electrode p and the bottom electrode q, and the memristor's internal state z. For example, if the memristor cell is reset ($= 0$), the state transition follows the top truth table in Figure 5.4. This behavior can be regarded as negative material implication ($z_n = p \to q = p \cdot \overline{q}$). If the memristor is set ($= 1$), it acts as reverse implication ($z_n = p + \overline{q}$). PLiM computer [96] regards this behavior as a 3-input majority gate function with an inverted input. Transformation of a memristor cell state z into a disjunctive normal form leads to the following equation using majority function M_3:

$$
\begin{aligned}
z_n &= (p \cdot \overline{q}) \cdot \overline{z} + (p + \overline{q}) \cdot z & (5.7) \\
&= p \cdot z + \overline{q} \cdot z + p \cdot \overline{q} & (5.8) \\
&= M_3(p, \overline{q}, z). & (5.9)
\end{aligned}
$$

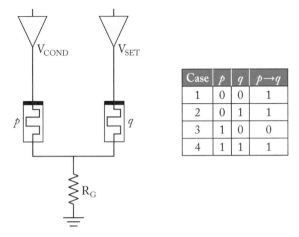

Figure 5.5: IMPLY's material implication logic and the truth table (adopted from [18]).

Using Resistive Majority function $(RM_3(p,q,z) := M_3(p,\bar{q},z))$, PLiM implements functionally complete operators and performs computation sequentially accessing a single bit in an array at a time. A compiler for PLiM computer [97] and a PLiM-based parallelized architecture using VLIW-like instruction set [98] are also proposed.

One of the limitations of the approaches above is that input data to the memristor-mapped gate has to be read out if it is stored in the memory. IMPLY [18] and MAGIC [17] propose gate mapping techniques without the need to read out the operands. IMPLY logic uses two memristor cells to implement material implication logic (Figure 5.5). Two inputs are stored in the cells p and q, and two different voltages, V_{COND} and V_{SET}, are applied to p and q, respectively, where $V_{COND} < V_{SET}$. If $p = 1$ (low resistance state), the voltage on the common terminal becomes nearly V_{COND} and the voltage on the memristor q is approximately $V_{SET} - V_{COND}$. Since this is small enough to maintain the logic state of q, q remains 0 if and only if $p = 1$ and $q = 0$. Otherwise, q remains 1 or turns into 1 due to the V_{SET} voltage applied to q. This scheme can naturally map to a memristor crossbar by regarding a bitline as the common terminal and a wordline driver as the voltage source. MAGIC [17] presents a scheme to map a NOR gate to the memristor crossbar using three cells (Figure 5.6). The output cell is initialized to 1, and two input cells are connected to V_{RESET}. Only when both of the input memristors are 0 (high resistance state), the voltage at the output cell is smaller than V_{RESET}, maintaining the initial value 1. Otherwise, it is reset to 0. MAGIC also propose other logic gate design (e.g., AND, NAND) but these cannot be directly mapped to crossbars. Siemon et al. [99] adopts a complementary resistive switch- (CSR-) based memristor cell to mitigate the effect of undesired current path (sneak path) and to improve the reliability of NOR and NAND gate mapping techniques.

Lebdeh et al. [100] propose a design for a stateful XNOR logic implementation on the ReRAM crossbar. Their design utilizes bipolar and unipolar memristors and allows cascading

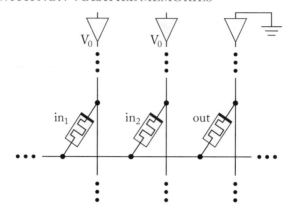

Figure 5.6: MAGIC's NOR logic and its array mapping (adopted from [17]).

the results by adding a single computing memristor and an additional cycle. While the bipolar memristor switches states depending on the magnitude and polarity of the applied voltage, the memristor in a unipolar mode needs suitable compliance current to perform a set operation. To reset the unipolar memristor, a higher level of current passes through the device and ruptures the conducting filaments. This switching mode of the memristor is used to compute the XNOR of two bipolar memristors, as the output voltage of the unipolar memristor changes only if there is a resistance gap between two inputs.

Gate-mapping techniques enable bulk logic operations with very low overhead. For some designs, the peripheral circuit need not add computation logic. However, they face similar challenges as the in-DRAM charge-sharing-based computing techniques. Given the fixed connectivity in the crossbar array, it is difficult to support a variety of logic operations in a small number of cycles. Thus, a single arithmetic operation composed of several logic operations can take hundreds to thousands of cycles to complete. While NVMs do not need to copy data thanks to non-destructive reads, they have limited endurance, slower operation frequency, and high write latency and energy. Thus, the gate-mapping techniques would be favorable to workloads that operate on massive read-only data with simple logic operations.

5.2   FLASH AND OTHER MEMORY TECHNOLOGIES

In addition to ReRAMs, other memory technologies, such as NAND/NOR flash memory, STT-MRAM, and PCM, have been considered for in-memory computing. This section will introduce some of the prospective research works for these technologies and their challenges. As a quick reference, Table 5.1 summarizes all NVM technologies and their characteristics. Note that the numbers in the table are obtained from most advanced circuit works, which often optimize the design for performance, but the numbers vary considerably depending on the array size, bit per cell, process technology, etc.

Table 5.1: Comparison of emerging memory technologies

	SRAM	DRAM	NAND Flash	NOR Flush	PCM	STT-MRAM	ReRAM
Cell area (F^2)	>100	6-10	<4 (3D)	10	4-20	6-50	2
Voltage (V)	<1	<1	<10	<10	<3	<2	<3
Read time	~1 ns	~10 ns	~10 μs	~50 ns	<10 ns	<10 ns	<10 ns
Write time	~1 ns	~10 ns	100 μs – 1 ms	10 μs – 1 ms	~50 ns	<10 ns	<10 ns
Write energy (J/bit)	~1 fJ	~10 fJ	~10 fJ	100 pJ	~10 pJ	~0.1 pJ	~0.1 fJ
Retention	N/A	~64 ms	>10 y	>10 y	>10 y	>10 y	>10 y
Endurance	>10^{16}	>10^{16}	>10^4	>10^5	$10^8 \sim 10^{15}$	>10^{15}	$10^8 \sim 10^{12}$
Multibit	1	1	>4	>4	>2	1	2–7
Non-volatility	No	No	Yes	Yes	Yes	Yes	Yes
F: Feature size of lithography							

5.2.1   FLASH MEMORIES

Flash memories are representative storage class memories that can provide both non-volatility and high density compared to conventional memories. The matured technology of NAND flash memory has enabled fast bulk storage systems, replacing HDD-based alternatives. Flash memories use floating gates which can capture and retain charges by their programming operation. Reprogram operation performs a bulk erase over a large *block* of memory cells (e.g., 512 rows of 16-KB page). Floating gates also allow multiple internal states by controlling the charge amount in the memory cell. Advanced MLC technology has enabled today's commercial flash memory to go beyond quad-level cells (16 states).

Near-Flash or In-Storage Computing

NAND flash has been pervasively presented as a fast storage memory for solid-state drives (SSDs) for its density and non-volatility, as well as its high-speed, low-power, and shock re-sistant features. SSDs communicate with a host system using standards including Serial Advanced Technology Attachment (Serial ATA or SATA) or NVM Express (NVMe). Because of the write-once and bulk-erase nature of NAND flash, SSDs have to properly buffer requests and manage blocks by performing garbage collection, interleaving multiple flash channels, etc. In addition, for better performance, reliability, and device lifetime, SSDs perform wear-leveling by performing indirect address mapping, which is enabled by an address translation in Flash Translation Layer (FTL). It is also in charge of managing the Error Correction Code (ECC) and bad block management. SSDs carry on the above-mentioned diverse tasks using firmware programs running on embedded cores based on ARM or MIPS architectures and using several GBs of DRAM to buffer data, operate FTL, and store address maps.

While near-disk computing was proposed in early works in the last century [101–106], these works faced difficulties providing enough performance gain to justify the cost because of the limitations of disk access speed and processing technology. With the advent of fast SSD devices, this concept has been re-examined, and several research works have leveraged embedded cores for near SSD computation. Generally, the bandwidth of high-end SSDs saturates at less than 4 GB/s due to bandwidth limitations on PCIe lanes. However, comparing the challenges to increase shared serial bus interfaces such as PCIe, increasing the internal bandwidth of SSD can be achieved more feasibly by introducing multiple flash memory channels and advanced fabrication processes. This potential for a larger internal bandwidth of SSD has attracted academic and industrial interest in near storage computing. These works offload compute kernels, including data analytics [107, 108], SQL queries [109, 110], operating system operations [111–113], key-value store [112, 114], object deserialization [115], graph traversal [116], image processing [108, 117], and MapReduce operations [118, 119], to SSD processors.

Notably, near storage programs need to be carefully designed as they make it difficult for current SSD processors to deliver a compelling performance. The performance of the embedded cores is still significantly lower than the host processor to make them cost-effective, so it is essential to consider the trade-off between the computation overhead associated with near SSD processing and the reduction in communication overhead to the host system. Summarizer [120] evaluated the trade-offs involved in communication and in-storage computing. The authors performed a design space exploration considering several ratios of internal SSD bandwidth and the host-to-SSD bandwidth and ratios of host computation power and SSD computation power, using several TPC-H database benchmarking queries. It is reported that processing the query workloads solely at the SSD leads to significant performance degradation because of the prolonged computation latency on the wimpy SSD controller core. They show that each application has an input-dependent sweet spot (thus difficult to be predicted) and collaborative computation between SSD and the host gives the best performance. They propose a framework that dynamically monitors the amount of workload at the SSD processor and selects the appropriate work division strategy between the host processor and the SSD processor. They also report, by using their dynamic load balancing scheme, in-SSD computation is more beneficial when internal bandwidth is higher than the external bandwidth and when a more powerful embedded core is employed, which can incentivize richer internal resources for future SSD platforms.

Near-storage computing has also attracted industrial interest, and several commercial products are being released. The Samsung SmartSSD computational storage drive [121] combines SSD and Xilinx's FPGA with a fast private data path between them, enabling efficient parallel computation at the SSD. The SmartSSD can be leveraged to provide data acceleration services (e.g., data compression, encryption, format transformation, etc.) and analytics acceleration services (e.g., database, image recognition, machine learning, etc.). ScaleFlux's SSD [122] has a GZIP compression/decompression engine and a customizable database engine accelerator. InnoGrit's Tacoma SSD [123] includes a mid-level Arm core and an implementation of

NDLA, an open-source inference accelerator from Nvidia. Interestingly, industrial efforts have been expended on integrating a customized/customizable ASIC into SSD, but not aggressively utilizing the computation power of SSD's embedded cores. We speculate this also reflects the challenges of gaining performance on the small embedded cores and properly selecting what to offload.

In-Flash Computing

In-memory computing in flash memories typically leverages voltage-current (IV) characteristics of a floating gate. This technique is explored in both NAND-based [124, 125] and NOR-based [126] approaches. For example, Wang et al. [124] activate pages in multiple blocks. Multiplication is performed in each bit-cell, and the currents accumulate on each bitline.

The benefits of flash-based in-memory computing include cell density, necessary gain, and energy consumption of the peripheral circuitry [126]. In particular, the high density of flash memories enables them to run a program with a large working dataset (e.g., large CNNs) without accessing an external memory. Also, flash is one of the most matured NVM technologies, making in-flash computing approachable. On the other hand, flash memories have several unique challenges. First, the read/write latency of flash memories is orders of magnitude larger than other volatile and non-volatile memories. Typical NAND flash takes a latency of more than 1–10 μs to read data, and it cannot be shortened for in-memory operations. Write (set) operation requires another 10× larger latency, and it super-linearly increases as the number of MLC states increases. Moreover, the asymmetric granularity of the set/reset operation makes content updates expensive. Albeit a single cell update, the block erasure necessitates copying and moving all the other data in the block. This makes it challenging to run a workload with a frequent update (e.g., neural network training). Second, the write endurance of the flash gate is lower than other memories, and its writable cycle count decreases nearly exponentially as the MLC level increases. Aged cells make them prone to read/write errors. Likewise, cell conductance decreases as the storage period increases, leading to retention error. A large array size for flash memory arrays potentially increases computation error due to the series channel resistance in the unselected layers [124]. Third, the exponential IV characteristics of a floating gate would make it challenging to control input signals from DAC. Mythic [125] approximates DAC using a digital value. NAND flash array performs accumulation of 8-bit cell contents ANDed by the input bit, and the peripheral circuit performs shift and add to make a final product.

5.2.2   STT-MRAM

Spin-Transfer Torque Magnetoresistive RAM (STT-MRAM) uses magnetic tunnel junctions (MTJs), whose magnetic layer can change its orientation using a spin-polarized current. Two magnetization orientations present different resistance states. STT-MRAM has a faster write speed, smaller write energy, and higher write endurance compared to other NVM devices. However, since it is challenging to implement MLC in STT-MRAM, in-memory computing on

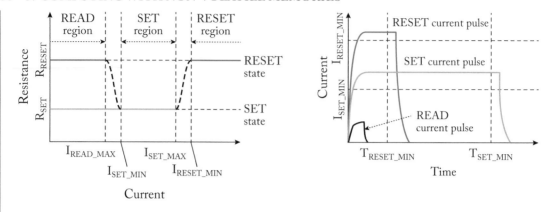

Figure 5.7: R-I curve (left) and current pulses (right) of PCM's operations.

STT-MRAM has focused chiefly on bitwise binary operations [127–129]. For example, STT-CiM [130] activates two rows simultaneously and senses current, which is one of the three distinctive values depending on the number of cells in the set state. Two sets of sense amplifiers using different reference currents produce AND and OR of two activated cells in each bitline at a time, and logic gates at the peripheral circuit combine these signals to produce XOR. In addition, STT-CiM supports addition by adding logic gates that compute the sum and carry, similar to Neural Cache [11], a digital in-SRAM computing work. Binary computation in STT-MRAM is also leveraged for binary convolutional neural networks [131]. While its application is limited due to the per-cell bit storage of STT-MRAM, its desirable cell endurance, read/write performance, and efficiency would make it a promising alternative to SRAM/DRAM-based approaches. Also, it is relatively mature compared to other emerging NVMs, and several works present the circuit and system level design.

5.2.3 PCM

Phase Change Memory (PCM) utilizes unique characteristics of chalcogenide glass or GeTe-Sb_2Te_2 superlattice. It uses the heat produced by the passage of electric current to quickly heat (600°C) and quench the glass and make it amorphous, or to hold it in its crystallization temperature (100–150°C) for some time and switch it to a crystalline state. The state change in the cell leads to different resistance. Also, by controlling the amount of material in the amorphous state, PCM can implement MLC. Using these favorable characteristics, PCM can be used for in-memory computation in a similar approach as ReRAM. There have been many circuit and device level research works that implement PCM-based in-memory computing for logical operations [88, 132], arithmetic [133, 134], and machine learning acceleration [135–139].

Despite the in-memory computing capability, PCM has several drawbacks that preclude PCM from becoming the best candidate for in-memory computing. Figure 5.7 shows the typical resistance vs. current (R-I) curve of a PCM cell and the current pulses needed for read, set, and reset operations. As shown in the figure, writes draw significantly more current and power than a read operation because of the heating operations, making write a slow and energy-consuming operation. According to Table 5.1, per-bit write energy is orders of magnitude larger than the other emerging NVM technologies. This necessitates higher voltage than the nominal supply voltage and charge pump circuits. Also, the current pulses for write operations are intense, and a PCM cell gets more unreliable as it experiences more writes. Additionally, MLC PCM suffers from long-term and short-term resistance drift [140], making it challenging to design even a 2-bit/cell main memory.

> **Pitfall** Analog in-memory dot-product computation can be easily extended for floating-point precision.
> This is not true because of the following reasons.
>
> - Floating-point addition and subtraction require normalization of mantissa bits if exponents are different. Therefore, after multiplications are performed in each row, the partial results need to agree with one exponent value for each column and mantissa is normalized according to the difference of the exponent. This shift amount is challenging to predict before calculating the product. Moreover, there is no straightforward mechanism to perform in-place normalization before summation at the bitlines. Thus, it is not possible to perform dot-product without writing the normalized products to different rows or arrays. This would result in a substantial throughput loss. In addition, since NVMs have limited write endurance and longer write time, it is not ideal for them to frequently store the intermediate calculation results.
>
> - If the entire mantissa bits are stored in one cell, the challenge of mantissa normalization seems less significant. Because analog values on cells sharing a bitline do not have to interact with the ones on another bitline, normalization only requires multiplication by 2^n ($n \in \mathbb{Z}$ is left shift amount), which DACs can encode. However, there is no straightforward way to perform matrix-vector multiplication with target exponents that can vary across columns. Also, because bits/cell is limited to at most 4 or 5, it is difficult to obtain a good dynamic range or accuracy. In-

creasing ADC's resolution is challenging because its area and power scale exponentially.

CHAPTER 6

Domain-Specific Accelerators

In this chapter, we provide the reader an overview of some of the novel domain-specific accelerator solutions that leverage in-/near-memory computing. We begin the chapter by discussing memory-centric acceleration approaches for machine learning. While machine learning has garnered the most interest from the architecture community, we also discuss some unique and interesting applications of in-/near-memory computing to other domains like automata processing, graphs, database, and genomics in this chapter. Besides these domains, near memory computing approaches have also been successfully used to improve memory security. For more details on the impact of smart memories in improving memory security, we refer the interested reader to the excellent discussion by Balasubramonian [141].

6.1 MACHINE LEARNING

Machine learning (ML) is one type of AI technique that is very effective in solving many real-world problems. In machine learning, a model is trained for some specific task and then the trained model is applied to some new inputs of the task and provides a solution. There are many types of ML models developed so far. Among them, the popular ones include Deep Neural Networks (DNN), Reinforcement Learning (RL), Support Vector Machine (SVM), and others. They have a wide range of real-world applications such as image classification, object detection, machine translation, game strategy, and customized recommendation. Many ML models are designed to solve otherwise hard problems and achieve a high accuracy.

To apply a ML model, there are two computation phases, training and inference. Training is the procedure to generate the parameters of the model, so that the model is best at the specific task. Inference is the procedure of applying the trained model to the new inputs and generating a result. The training phase is usually heavier in computation than inference. During training (note: supervised training), some inputs are fed to the model for a normal prediction and the parameters are then adjusted based on the difference of the output of the model and the ground truth results.

DNN is a general term for one family of ML models. DNN has many layers stacked together, and the output of one layer becomes the input of the next layer. One DNN layer usually includes a linear transformation on the input, and then is followed by a nonlinear function. There are trainable parameters in the DNN layers, which are also called weights. Usually, the linear transformation step involves a large number of weights and dominates the computation in inference. The most common DNN models are Multi-Layer Perceptron (MLP), Convolutional

Neural Network (CNN), and Recurrent Neural Network (RNN). In MLP, the linear transformation is matrix-vector multiplication where the vector is the input and the weights form the matrix. CNN is mostly composed of convolutional layers, where the linear transformation is the convolution operation. In such a convolution, the input and output are both 3D tensors and the weights are an array of 3D tensors which are also called filters. In RNN, the linear transformation is matrix-vector multiplication similar to the MLP. The difference is that the RNN takes a sequence as the input. The input vector in RNN includes the output vector from the previous element in the input sequence, of the same layer. In summary, the key operations in DNN inference include convolution and matrix-vector multiplication.

Many recent works have proposed to accelerate ML applications with in-memory processing [11, 38, 39, 53, 66–69, 73, 77, 78, 86, 87, 91–93, 142–157]. There are a few reasons that in-/near-memory computing is a good fit for accelerating ML. First, ML applications usually have a high degree of parallelism as the computation involves large matrices and vectors. The parallelism can be well exploited by the vector-processor style architecture of in-memory processing. Second, in the inference stage of a ML application, the model parameters are usually fixed and thus can be pinned to the memory, while the inputs are being streamed in. This is especially a desired behavior of in-memory processing as one of the operators (weights) is stationary and only the movement of the inputs needs to be orchestrated.

The in-memory processing systems mainly target the key operations, matrix-vector multiplication (MVM), and convolution in the ML applications, as they are the most time-consuming operations in many ML models. The MVM and convolution usually can be broken down as a series of multiplication-accumulation (MAC) operations and they are mapped to the massively parallel in-memory processing units. There are usually separate functional units for the nonlinear and other element-wise operations, which are important components to ensure the effectiveness and accuracy of ML systems, but take only a small amount of time.

In the next sections, in-/near-memory processing architectures for ML are described in detail, classified by the memory medium type. For a more detailed coverage of accelerators for ML applications (especially DNNs), please refer to [158].

6.1.1   ACCELERATING ML WITH NON-VOLATILE MEMORIES

In Chapter 5, the properties and basic arithmetic operations with NVM were discussed. The ReRAM, due to its highly efficient MAC operation, has been widely adopted in ML applications [77, 78, 91–93, 142–147]. Next, we will discuss how a deep learning application is mapped to in-NVM computation operations, with the case study on one typical design, ISAAC [78], which accelerates CNN inference on ReRAM.

Hardware Architecture   First, we introduce the overall hardware architecture of ISAAC. At a high level, the ISAAC chip consists of multiple tiles, which are connected by a c-mesh interconnect. Each tile has multiple In-Situ MAC (IMA) units, eDRAM modules (as buffers for inputs), as well as dedicated computation units for computing the sigmoid function and max-pooling.

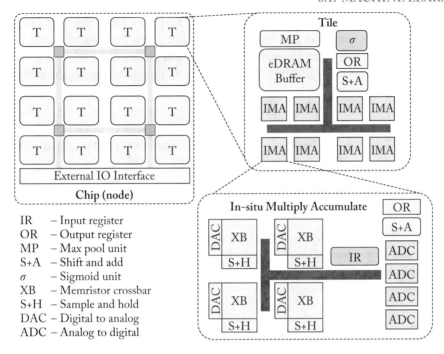

Figure 6.1: The overall architecture of ISAAC [78].

The modules within a tile are connected by the bus. An IMA unit contains essential modules for massive MAC operations with ReRAM; there are multiple ReRAM crossbars, DAC and ADC units, input/output registers, and shift-and-add units (for merging the results of different bit-positions from different columns). The architecture of ISAAC is summarized in Figure 6.1.

Execution Model   As ISAAC aims to minimize the costly read/write of the weights to the ReRAM, the weights of the entire model are partitioned by layer and loaded to different tiles at the preprocessing step. For example, tiles 1–4 may be assigned to the first layer of the CNN, tiles 5–12 to the second layer, and so on. The computation for the different layers is then done in a pipelined manner. In the above example, the first layers' results are transferred from tiles 1–4 to tiles 5–12 for the computation of the next layer, and after that tiles 1–4 can work on a new batch of inputs. ISAAC also increases the total utilization of the ReRAM array, by making each layer's arrays work on the partial results produced by the previous layer.

In the ReRAM crossbar, the weights from a filter are mapped to the multiple cells in the same column, so the matched inputs are applied at the rows to calculate the MAC. The weights from different filters are mapped to different columns of the ReRAM arrays to calculate multiple outputs simultaneously with the same set of inputs. The inputs are stored in the eDRAM, and sent to ReRAM sequentially for computing the outputs at different positions sequentially.

Other Works with ReRAM   PipeLayer [92] is a ReRAM-based in-memory computing architecture for CNN training. Within a ReRAM crossbar array, the data mapping is similar to ISAAC: weights from one filter are mapped to a column (bitline). PipeLayer introduces one pipelining strategy for training: for the entire forward and backward passes, the computation of each layer on one image is treated as a stage in the pipeline, and multiple images are processed for different layers at the same time. Further, some memory arrays are used for buffering the intermediate results necessary for the sophisticated data dependency in CNN training.

FPSA [147] proposes a flexible in-ReRAM accelerator for DNN, with a reconfigurable routing architecture and place-and-route tools. FloatPIM [91] supports floating-point values with fully digital in-memory computation, and therefore enables training with high numerical precision. It also optimizes data transfer across memory blocks by inserting switches among the blocks.

There are also works on fabricated chips for ReRAM-based in-memory DNN acceleration [86, 87]. They use a similar mapping strategy for the weights and inputs, but evaluate with a lower precision for the weights (2–3 bits) and inputs (1–2 bits), and smaller DNN models or datasets (e.g., MNIST [159]).

Other Types of NVM   Beyond ReRAM, there are other types of non-volatile memories that have been proposed for ML acceleration, including Phase Change Memory (PCM) [148], Spin-Transfer Torque Magnetic RAM (STT-MRAM) [149], and Ferroelectric RAM (FeRAM) [150]. There are trade-offs among the different NVM types (details in Table 5.1). For example, STT-MRAM has been leveraged mostly for ML models with binary weights or activations, as it is challenging to store multiple bits in a cell in STT-MRAM. PCM-based architecture incurs high energy consumption for training tasks, as training requires a lot of data writing and the writing energy for PCM is relatively high.

6.1.2   ACCELERATING ML WITH IN-SRAM COMPUTING

Hardware Architecture   Chapter 4 discussed how to perform in-SRAM computation with bit-serial algorithms and transposed data layout. Neural Cache [11] repurposes the last level cache of general purpose CPUs as in-SRAM computing units, and maps the computation of CNN inference onto them. Please refer to Section 4.1 for the background on the structure of CPU last level cache. Most of the SRAM arrays in the cache are used for in-memory computation, and a small portion are used for buffering the outputs of a layer. The interconnect system for the cache is reused for in-SRAM computation. The data are transferred across the set-associative ways in the intra-slice data bus and across multiple slices on the ring-structured inter-slice data bus. The CPU is responsible for initiating the computation in the cache. The data may also be loaded from the main memory to the cache when necessary.

Execution Model   In Neural Cache, the different layers of a CNN are computed sequentially, to maximize the resource utilization within each layer. The convolutional, fully connected, and pooling layers are all computed with the bitline ALUs in SRAM.

Below is how the computation for one convolutional layer is executed. The core operation of convolution (and also many other ML models) is multiplication-accumulation (MAC), which involves the multiplication of weights of a 3D filter and the corresponding inputs and accumulation of all the products. The 3D filter is mapped to one 2D region in the SRAM array, and the corresponding inputs are mapped to the 2D region with the same set of the bitlines so that the MAC is performed with the bitline ALUs in parallel. Adding up the MAC results at multiple bitlines in an array is performed by a series of inter-bitline copy operations and then addition. The entire set of 3D filters are mapped to the neighboring SRAM arrays and the inputs are broadcasted to them. Then all the filters are further replicated and mapped to the rest of the bitline ALUs in SRAM to calculate even more outputs simultaneously. The weights are loaded from DRAM to all the cache slices via the ring bus, and the inputs are broadcasted within a cache slice with the intra-slice data bus.

In contrast to the digital in-memory computing approach in Neural Cache, some other works apply analog computing in SRAM [66–69, 73, 151–153]. As they are using a similar mechanism for computing the MAC on the bitline as the ReRAM (weights are stored in the memory cell and inputs are sent in to all wordlines for MAC), the data mapping of ML applications is similar to the ReRAM-based solutions.

6.1.3   IN/NEAR-MEMORY ML WITH DRAM

There are both near-DRAM and in-DRAM computing works that target ML applications [38, 39, 53, 154–156].

Many works on near-DRAM computing for ML are based on 3D stacked memories (general structure discussed in Section 3.2). The main consideration in mapping the ML model is to reduce the data movement, especially in the Network-on-Chip in the logic layer. This is equivalent to reduced data transfer across different vaults. Neurocube [38], for example, partitions the input image into smaller tiles and maps them to different vaults, so that the convolution filter in each vault only needs at most a small number of edge pixels from other vaults. For a fully connected layer, the inputs are duplicated to multiple vaults, where each input vector is multiplied with a partial weight matrix. Tetris [39] further optimizes the data mapping by proposing a hybrid work partitioning scheme which considers the total amount of memory access, in addition to inter-vault data access. Tetris also follows the row-stationary dataflow [160] for convolution.

An example of in-DRAM computing architecture for DNNs is DRISA [53]. Its underlying triple-row activation computing technique is discussed in Section 3.4. At a high level, it builds a vector processing architecture based on activating multiple DRAM arrays simultaneously. In the data mapping for CNN, the weights from different filters are mapped to the different bitlines in an array and stored there. The inputs are then broadcasted to all the bitlines. When one filter cannot be mapped to a single array, the results from all arrays are added up.

6.2   AUTOMATA PROCESSING

In this section, we demonstrate the applicability of in-DRAM computing techniques to accelerate another important application, automata processing. Finite State Automata (FSA) or Finite State Machines (FSMs) are a powerful computation model for pattern matching. Many end-to-end applications in data analytics and data mining, web browsing, network security, and bioinformatics make use of finite state machines to efficiently implement common pattern matching routines.

FSM computation also accounts for a significant portion of the overall runtime in many applications. For instance, FSM-like parsing activities can account for up to 40% of the loading time of many web pages in web browsers [161] and 30–62% of the total runtime of DNA motif search in tools like Weeder [162]. However, since FSM computation is *embarrassingly sequential*, it is hard to accelerate [163]. Regardless of how well other parts of these applications are parallelized, without acceleration of these sequential FSM operations [164], only small improvements in end-to-end application performance can be obtained (Amdahl's Law).

We begin this section by providing a refresher on FSMs in Section 6.2.1. Section 6.2.2 highlights some of the limitations of automata processing on conventional CPUs/GPUs such as limited parallelism, poor branch behavior, and limited memory-bandwidth. Micron's Automata Processor (AP) overcomes these limitations by embedding FSM computation inside a DRAM chip, effectively transforming DRAM arrays into massively parallel FSM computation units with access to >800 GB/s of internal DRAM bandwidth. Section 6.2.3 discusses the Micron's Automata Processor architecture in detail. We end this section by discussing recent innovations in programming models, architectural optimizations for the Automata Processor as well as recent work targeting other memory substrates (e.g., SRAM) for automata processing.

6.2.1   FINITE STATE MACHINE (FSM) PRIMER

A finite state machine consists of a finite number of states with transition rules governing the transitions between these states. The transition rules are defined over a finite input alphabet. A finite state machine detects patterns in the input stream by repeated application of the transition rules to every active state. As discussed earlier, FSM computation is inherently sequential. Before processing the next input symbol in the stream, all state transitions corresponding to the previous input symbol from all active states in the finite state machine need to be completed.

A deterministic finite state automaton (or DFA) has atmost one active state and one valid state transition on an input symbol. In contrast, a non-deterministic finite state automaton (or NFA) can have multiple active states and multiple state transitions on the same input symbol. While NFAs and DFAs have the same expressive power, NFAs are much more compact.

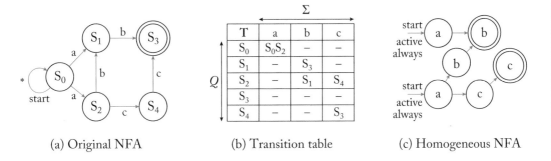

| (a) Original NFA | (b) Transition table | (c) Homogeneous NFA |

Figure 6.2: (a) Original NFA to detect patterns ab, abb, and acc. Start states are indicated as start and reporting states are shown using double circles. $\Sigma = \{a, b, c\}$ is the input alphabet. (b) Transition table for the NFA. (c) Equivalent homogeneous NFA representation for the NFA shown in (a).

Table 6.1: Storage complexity and per-symbol computation complexity of an NFA with n states and its equivalent DFA. Σ is the input alphabet.

	Processing Complexity	Storage
DFA	$\mathcal{O}(1)$	$\mathcal{O}(\Sigma^n)$
NFA	$\mathcal{O}(n^2)$	$\mathcal{O}(n)$

An NFA is formally described by a quintuple $\langle Q, \Sigma, \delta, q_0, F \rangle$, where Q is a set of states in the automaton, Σ is the input symbol alphabet, q_0 is the set of start states and F is the set of *reporting* or *accepting* states. The transition function $\delta(Q, \alpha)$ defines the set of states reached by Q on input symbol α. If the automaton activates the reporting state, then the input symbol position in the stream and the reporting state identifier are reported. Figure 6.2a shows an example automaton to detect patterns ab, abb, and acc in the input stream. The transition table for the same automaton containing $|Q| \times |\Sigma|$ transition rules is shown in Figure 6.2b.

Table 6.1 shows the processing complexity per input symbol and storage requirements of an NFA with n states and its corresponding DFA. Since NFA can have up to n active states in any symbol cycle, and up to $n - 1$ transitions to other states, it has a computation requirement that scales as $\mathcal{O}(n^2)$ per input symbol. In contrast, in a DFA with atmost one active state, only one state transition rule needs to be fetched from memory per input symbol. As a result, only a constant amount of work is performed per input symbol. The low computational requirement of a DFA has made it an attractive option for Von Neumann architectures like CPUs and GPUs. However, a DFA has large storage requirements that grow exponentially with n, since a DFA may require up to $|\Sigma|$ states for every state in the original NFA.

A homogeneous NFA imposes some restrictions on the original NFA such that all incoming transitions to a state occur on the same input symbol. This restriction allows each state to be labeled with the corresponding input symbol that it is required to match. This is achieved by adding additional states to the original NFA. Homogeneous NFAs have equivalent representative power to their original NFA counterparts [165], but enable hardware-friendly acceleration. In Figure 6.2c we can see that the state S_3 with two incoming transitions on different input symbols is split into two different reporting states labeled b and c.

6.2.2   COMPUTE-CENTRIC FSM PROCESSING

NFA computation involves scanning a stream of input symbols for a set of pre-determined patterns, by computing the set of active states that match the current input symbol and looking up the transition function to determine the set of next active states for each input symbol.

Conventional CPUs perform poorly on automata processing. NFA implementations designed to run on CPUs are typically implemented in two ways. First, using if-else or switch-case nests to encode the transition function. However, these branches are data-dependent and hard to predict leading to poor branch behavior and low performance. Second, storing the complete transition function as a table typically in off-chip memory. Since there is a lookup into this table for every active state and input symbol, there is significant amount of time spent in data movement across the memory hierarchy. Furthermore, transition table accesses are irregular. There is also very little computation performed per input symbol (i.e., computing the next active state).

Although NFA computation is inherently sequential, there have been efforts to exploit other sources of parallelism available in automata processing: (1) input stream-level parallelism, which involves processing multiple input streams in parallel on the same NFA; (2) NFA-level parallelism, which processes the same input stream on multiple NFA (similar to the Multiple Instruction Single Data paradigm); and (3) NFA state-level parallelism, which involves parallel exploration of all valid paths in the NFA, starting from each active state. However, modern multi-core CPUs have limited number of threads and can process only a few input streams or NFA patterns or NFA paths in parallel. Even the available parallelism cannot be often fully leveraged for large NFAs with many active states because of memory bandwidth limitations. In contrast, GPGPUs have significant available parallelism and high memory bandwidth. But they suffer from significant memory divergence arising from irregular memory accesses to the transition table stored in global memory. Also, GPU implementations that use a static mapping of NFA states to threads suffer from load imbalance and low hardware utilization due to idle threads.

6.2.3   MEMORY-CENTRIC FSM PROCESSING

As discussed in the previous section, memory bandwidth is one of the main limiting factors for many automata processing problems. We now explain a canonical memory-centric automata processing architecture, Micron's AP [9] which has been hugely successful in accelerating several applications involving automata processing, for example, database entity resolution by 434× [166] and motif search by 201× [167].

AP overcomes the memory bandwidth limitation of automata processing by directly implementing NFA states and state-machine logic inside a DRAM memory chip. The memory arrays provide access to more than 800 GB/s of internal bandwidth to the state-machine logic, far greater than available pin bandwidth to CPUs. AP has been highly successful at automata processing for three main reasons. First, the AP leverages the massive bit-level parallelism of DRAM memory arrays to traverse multiple active paths in the NFA in parallel. The AP supports up to 48 K state-transitions in a single cycle. It does this by mapping NFA states to DRAM columns and implementing a custom routing matrix to support highly efficient state-transition near DRAM memory arrays. Second, the AP is highly energy efficient because it eliminates the costly data movement between the CPU and memory and reduces Von Neumann instruction processing overheads significantly. Third, the AP has deterministic throughput irrespective of the complexity of the underlying automata and can efficiently execute large NFA with several thousand patterns.

Concept   Figure 6.3 provides a conceptual overview of NFA execution on a memory-centric architecture like the AP. The automaton detects patterns ab, abb and acc in the input stream. The start states are labeled as start and in the example automaton are designated to be always-active, i.e., these states are activated for every input symbol. Each input symbol is processed in two steps. First, in the **state-match** step all states in the NFA that have the same label as the input symbol are determined. We refer to these as the set of matching states. However, only a subset of these matching states may be active for a given symbol cycle. These are obtained by intersecting the set of matching states with the set of currently active states. We refer to the resulting set obtained from intersection as the set of enabled states. Next, in the **state-transition** step, the destination states for each of the enabled states are determined. These destination states become the set of active states for the next symbol cycle.

Micron's Automata Processor   Micron's Automata Processor is a DRAM-based hardware accelerator that accelerates NFA execution. Besides supporting regular NFA, it also includes counter and boolean elements to support a large class of automata applications, beyond regular expressions. The AP fits in a regular DIMM slot and can be interfaced with the host CPU via the DDR/PCIe interface. The initial AP prototype was implemented in a 50-nm technology node and runs at 133.33 MHz. A single AP board provides a throughput of 1 Gbps, when processing 8-bit input symbols.

In Micron's AP, the columns of the DRAM memory array are re-purposed to store homogeneous NFA states also called State Transition Elements (or STEs) as shown in Figure 6.4. Each STE is associated with a 256-bit symbol recognition memory. This memory stores the 8-bit input symbol/symbols that the STE is programmed to recognize using one-hot encoding as

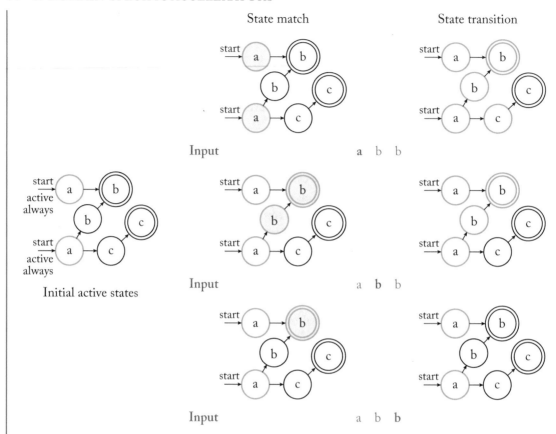

Figure 6.3: Example showing parallelism sources in NFA processing that can be leveraged by memory-centric architectures. Each input symbol is processed in two steps: state-match and state-transition. Multiple active states and their corresponding edges can be explored in parallel.

shown in Figure 6.4a. For example, an STE programmed to recognize the ASCII input symbol *b*, would only have the bit position in row 98 set. To keep routing complexity tractable, each STE is provisioned to have up to 16 incoming transitions from other STEs (16-bit STE enable input in Figure 6.4a).

Furthermore, the row address is also repurposed to send the 8-bit input symbol to the DRAM arrays after passing through the row decoder as shown in Figure 6.4a. Each cycle, the same input symbol is broadcast to all memory arrays as the row address. If an STE in a memory array has a bit set corresponding to the row address, then the STE label matches the input symbol. As a result, a simple DRAM row read operation is sufficient to implement **state-match** functionality in a massively parallel fashion. Each STE also includes an active state bit to indicate if it is active in a particular symbol cycle. A logical "AND" of the state-match results with

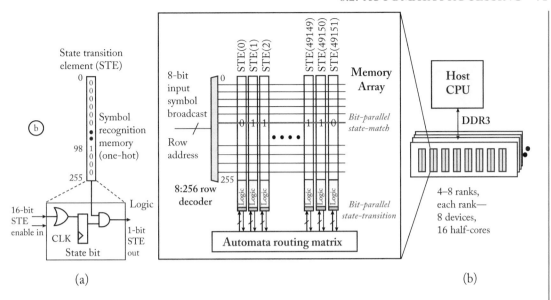

Figure 6.4: (a) State-transition element in the AP with symbol recognition memory and next-state activation logic. (b) AP interfaced to host CPU via DDR3. AP leverages bit-level parallelism of memory arrays for massively parallel NFA processing.

the active state bit vector gives the set of enabled states. To implement the **state-transition** functionality, the bit-vector of enabled states is made to pass through a custom hierarchical routing matrix consisting of several switches and buffers, which encode the transition function for the NFA. The output from these switches is a bit-vector of active states for the next symbol cycle. These update the local active state bits for each STE in parallel. The switches that are part of the routing matrix are programmed prior to execution as part of a compilation step. Reconfiguring these switches is time consuming when compared to only updating the STE memory to match a new set of input symbols.

Due to physical placement and routing constraints, each AP device is organized hierarchically into half-cores, blocks, rows and states as shown in Figure 6.5. States mapped to different half-cores have no interconnections between them. Each AP device has a TDP of 4 W. Power gating circuitry is included at each block to disable blocks with no active states. The AP supports designating few NFA states as `reporting states`. The current generation of AP supports 6 reporting regions, with up to 1024 states per region. Reporting events are communicated to the host CPU by writing to a designated output event buffer in DRAM, which is later post-processed in the host. Each entry contains an identifier for the reporting state and the input symbol offset in the stream causing the report. The AP also includes a 512-entry state-vector cache to support incremental automata processing. This cache allows the AP to context switch

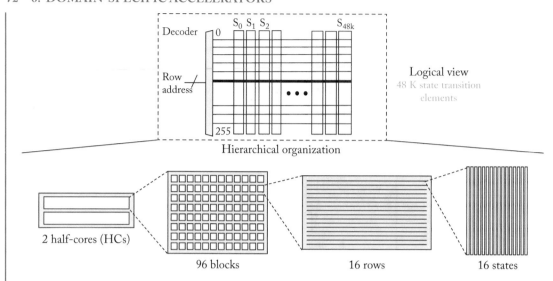

Figure 6.5: Logical view and detailed hierarchical organization of state transition elements in one Micron's automata processor device.

between different input streams for multiple users, akin to register checkpointing in traditional CPUs. Context switching on the AP is fast, however, taking only three cycles.

Programming Support   One of the primary modes of programming the AP is by defining NFAs in an XML-like language format, called ANML (Automata Network Markup Language) developed by Micron. The ANML specification is compiled into a binary image, which is later placed-and-routed on the AP device using Micron's proprietary SDK. While ANML offers rich functionality, programming in ANML has been found to be tedious, verbose and error-prone, akin to CPU assembly programming. Another option is to specify patterns using regular expressions, but it is also non-intuitive and error-prone. RAPID [168, 169] is a recently proposed C-like language designed to address the above limitations. It includes parallel control structures customized for parallel pattern matching against an input stream. To expand support beyond AP, the authors also propose a JSON-based unified format, MNRL for designing and manipulating NFA for different computing architectures like FPGAs [170]. To ease debugging of RAPID programs, Casias et al. [171] leverage per-cycle automata state information stored in the AP state-vector cache to support breakpoints on input data.

Architectural Extensions   There have also been several recent architectural optimizations proposed to improve the performance and utilization of the AP. Wadden et al. [172] observe that the AP has limited I/O pin bandwidth and is optimized for automata with infrequent but dense reports (i.e., many states report on the same cycle). Automata benchmarks with frequent reporting can experience up to 46× slowdown. To overcome this limitation, the authors propose software transformations and hardware extensions like report aggregation and report queueing to improve I/O bandwidth utilization. Since a single AP device can only support 48 K states, Liu et al. [173] aim to reduce the overheads of reconfiguring the AP for large NFAs by partitioning them into hot and cold states. By spending most of the AP resources to execute transitions from hot states, a 2.1× speedup over baseline AP is observed. Subramaniyan et al. [174] break the sequential processing bottleneck of the AP for a single input stream by proposing low-cost extensions to the AP to support enumerative parallelization [175]. Enumeration is made feasible by terminating enumeration paths in the NFA that are inactive or have converged. For further details, we refer the interested reader to the excellent survey by Mittal [176], which provides a more extensive discussion of architectural extensions to the AP.

Leveraging Other Memory Substrates   Several recent works have demonstrated the potential of other memory technologies like SRAM [177, 178], eDRAM [178], and ReRAM [179] to accelerate automata processing. Micron's Automata Processor spends a significant amount of die area implementing the routing matrix in a lower technology DRAM node. The AP includes only about 200 Mbit of physical DRAM, while a regular DRAM chip can store 2 Gbit of data in the same area. Subramaniyan et al. [177] show that faster and energy-efficient SRAM-based last-level caches can be repurposed to accelerate automata processing by 9–15×. There have also been efforts to overcome the 8-bit fixed symbol processing limitation of the spatial memory-centric automata processing architectures. Sadredini et al. [180] improve the utilization of state-matching resources by enabling processing of multiple 4-bit symbols per cycle. To decouple the bit-width used in the NFA (given by alphabet size) from the bit-width used in the memory-centric hardware implementation (number of rows in memory array), Sadredini et al. [181] propose a automata compiler to enable arbitrary bit-width processing on spatial reconfigurable architectures like the AP. Sadredini et al. [178] also observe that full crossbars for state-transition are overprovisioned for many real-world NFA and propose a reduced crossbar based interconnect to improve area-efficiency for state-transition.

6.2.4   AUTOMATA PROCESSING SUMMARY

Memory-centric automata processing architectures like the Micron's AP have leveraged the massive bit-level parallelism of DRAM arrays to accelerate a number of pattern matching workloads. They have enabled solutions to certain string matching problems previously unsolvable by conventional von Neumann CPUs [167]. AP provides deterministic throughput (1 Gbps per device) irrespective of automata complexity, supports a large class of automata applications, supports incremental automata processing, and demonstrates linear performance scaling with increasing number of ranks. Other potential acceleration candidates and architectural extensions to the AP are being actively explored both in academia and industry (e.g., Natural Intelligence Systems).

6.3   GRAPHS

A graph formalizes relationships between objects. For example, social graphs, web graphs, and citation graphs are some of the representative graphs in the real world. These graphs are often composed of millions to billions of vertices and edges connecting them. Graphs are often represented using an adjacency matrix. Given the number of vertices $|V|$, the adjacency matrix forms a square matrix with $|V|^2$ elements where element (i, j) indicates whether vertices v_i and v_j are adjacent. Real-world graphs are often sparse in the connectivity of the entities, which produces a sparse adjacent matrix with a significantly large number of null (zero) elements. Therefore, graphs are usually provided in a compressed format, such as compressed sparse row (CSR) and coordinate list (COO).

The complexity and diversity of graph workloads lead to various computation behaviors and demands. Modern graph applications can generally be classified into three categories [182].

1. **Graph Traversal**: The majority of computation occurs during graph traversal, jumping from a vertex to its neighbors. Due to the scattered and sparse nature of graph connectivity, its traversal incurs a large number of irregular memory access. Example applications: breath-first search (BFS), single source shortest path (SSSP), PageRank.

2. **Rich Property**: Vertices are associated with rich properties (e.g., large stochastic tables). Computation processes the properties and is numeric heavy, showing similar behavior to conventional applications. Example applications: Gibbs inference, Bayesian network.

3. **Dynamic Graph**: Graph structure dynamically changes over time. While they exhibit memory-intensive behavior and irregular memory access patterns similar to Graph Traversal applications, their computation is more diverse because of dynamic memory footprint and write-heavy memory access.

To understand the fundamental behavior of popular graph applications, we focus on applications in the Graph Traversal category.

Challenges   One of the challenges of graph computing is irregular memory access. Traversing CSR formatted data inherently incurs indirections, necessitating multiple memory accesses to fetch each non-zero element in the sparse matrix. Furthermore, the random traversal order leads to a large number of cache misses. Likewise, computation associated with each data fetch is relatively lightweight, making the applications memory bound.

The random unpredictable nature of graph traversal order also incurs a large memory footprint, which brings a challenging problem for multi-node near- or in-memory computing, where working data can span across different memory modules. In such a situation, remote memory accesses need to go through narrow-bandwidth off-chip links, and they must also be aligned with address translation, cache coherence, and inter-core synchronization.

Moreover, the random scattered access patterns make in-memory computing less efficient. While in-memory computing on a memory array generally requires operands to be arranged in a designated location to perform computation and exploit parallelism, this scheme cannot be directly applied to a compressed storage format of a sparse matrix. To expose the computation models of in-memory computing, sparse matrices need to be decompressed to the dense matrix format, reinserting null elements eliminated by the compression. Decompression leads to the following inefficiencies. First, it triggers data movement in the memory, paying the cost for bandwidth, energy, and cell write count for additional write cycles. Second, because of the sparsity, computation density per array becomes low, undermining the dense compute capability of a highly parallel architecture. Third, while there is a locality in edge access (e.g., identifying adjacent vertices in the CSR format), access patterns of vertices are generally irregular, causing only a subset of decompressed data to be utilized, given a general vertex-centric scheme is used. This not only reduces computation density in a decompressed matrix, but also results in repetitive decompression to access vertices in the dense format across different timeframes.

While the irregular memory access generates enormous bandwidth demands for the memory subsystem during execution, graph applications generally present abundant data-level parallelism. For example, BFS uses a set of frontier vertices to reference the child vertices of a node for a given iteration, and the child reference process for the vertices in the set can be parallelized. Near-memory computing for graph applications leverages the internal memory bandwidth of stacked memories to minimize traversals of off-chip links, taking full advantage of the parallel processing units near the memory.

Near-Memory Graph Processing   Many near-memory solutions for graph application adopt fully programmable cores, such as in-order ARM Cortex-A series. For example, Tesseract [37] has 32 ARM Cortex-A5 processors with floating-point unit (FPU) per HMC, whose area is about 9.6% of 8 Gb DRAM die area. Tesseract supports inter-HMC-node communication through message passing. Message passing offers a simple interface but can avoid cache coherence issues of Tesseract cores and locks to guarantee atomic updates of shared data. Near-memory computation can be triggered by blocking and non-blocking remote function call on message passing. Tesseract also proposes a prefetching scheme of a remote node through message communication and a simple programming interface. One challenge of using the message passing scheme could be proper data distribution and application verification. Programmers are in charge of writing a correct program to eliminate potential data races and deadlocks, communicating messages with correct nodes, and flushing caches if any update is made by CPU, etc. Also, while there is randomness in vertex access, real-world graphs are often skewed in their non-zero distribution, making it challenging to load balance across nodes. There are several pieces of architectural work based on Tesseract [183–185]. GraphP [183] takes data organization as a first-order design consideration and proposes a hardware-software co-designed system. GraphQ [184] enables static and structured communication with runtime and architecture co-design. GraphH [185] integrates SRAM-based on-chip vertex buffers to eliminate local bandwidth degradation.

Gao et al. [33] also adopt fully programmable in-order ARM Cortex-A cores, but support virtual address space through OS-managed paging. Each core contains a 16-entry TLB for address translation, using 2 MB pages to minimize TLB misses. TLB misses are handled by the OS on the host, similar to an IOMMU. Since a memory node belongs to a unified physical address space in the memory system, read/write access to a remote node can be routed by a network-on-chip router on the logic die. Cache coherence for the near-memory core is managed by a simple software-assisted coherence model. The software interface is provided by a simple API which calls a runtime software providing runtime services such as synchronization and communication. This runtime API can be driven by middleware such as MapReduce and GraphLab framework.

GraphPIM [182] observes that atomic access to the graph property is the chief cause of inefficient execution of graph workloads. The data components used in graph applications can be classified into three categories: (1) Metadata (local variables and task queues, which are small and cache-friendly); (2) Graph structure (the data structure that contains the compressed adjacent matrix used to retrieve the neighbor vertices); and (3) Graph Property (the data associated with each vertex. Due to the irregular access pattern to vertices and the large size of the graph, it can rarely be captured by caches). They identify that the computation on the property data is a simple read-modify-write (RMW) atomic operation, and target offloading the atomic operations to a near-memory processing unit. While fully programmable near-memory cores offer great flexibility, they introduce non-negligible hardware complexity. Instead, GraphPIM uses atomic units in the HMC, and adds a PIM offloading unit (POU) to the host core, which modifies the data path of memory instruction considering the target memory address and instruction. The use of the fixed-function unit minimizes the hardware complexity, and POU allows offloading without any modification of application and ISA.

In-Memory Graph Processing   GraphR [186] presents an in-memory accelerator based on ReRAM crossbars. They observe that many graph algorithms can be expressed in an iterative procedure of sparse matrix-vector multiplication (SpMV) and accelerate the computation using ReRAM. Since the compressed format cannot be directly used for in-memory computation, they convert a submatrix of compressed adjacency matrix to a dense format and load it to a ReRAM crossbar. While submatrix with all zeros can be skipped, sparsity in the input graph can cause the above-mentioned computation density issues. Real-world graphs have a variety of non-zero distribution, and often it has significant skewness (non-uniformity), which increases the chance to find relatively dense submatrix and all-zero submatrix. However, whether such an opportunity can be exploited is heavily dependent on the input graph.

Discussion   Overall, the sparse nature of graph data and irregularity in memory access make a challenging situation for memory subsystems. Near-memory acceleration is an active field of study, and its interest lies in what/how to offload graph applications, how to support efficient communication, and where to perform computation. Subdividing work is also an important question to discuss. Generating a submatrix from a compressed format incurs preprocessing overhead, which cannot be ignored for a type of workload where graph structure is dynamically generated or is subject to change during execution. In-memory computing requires decompression into a dense format. Thus, it is challenging to reap the benefit of the reduction of data movement from the data's original location. Potentially, in-memory computing can be performant if it supports more flexible interconnects within an array for in-place computing using a compressed format or supports a hybrid array organization (compute array and memory array) to exploit parallelism in an array.

6.4 DATABASE

Database applications can roughly be categorized into two groups: a relational database management system (RDBMS) and NoSQL. RDBMS manages a database based on the relational model of data and processes SQL queries. On the other hand, NoSQL does not depend on the fixed schema of the relational model and the *join* operations, offering good horizontal scalability with a distributed data store. Key-value store (KVS) is one of the representative NoSQL databases.

RDBMS processes SQL queries by first making an optimized query plan. It is an execution flow of database operators (e.g., scan, group by, join, and sort), and appropriate algorithms for each operator are selected based on the expected table size and pre-built index table. Because of the large size of the database, table data is typically stored in storage drives such as HDD, and for faster lookups, RDBMS often creates index tables using B-trees or hash tables. Index tables are usually stored in the main memory. Database algorithms using index tables such as index scans often cause random access to the tables. This implies, if the result selectivity (i.e., the ratio of table entries returned after a scan) is low, the index scan minimizes memory access, while a higher selectivity would result in performance degradation compared to a full (sequential) scan due to random access. Moreover, a query plan is generally composed of several data manipulation operations that incur data movement. For example, suppose a hash join is performed on scanned data. In that case, a table is scanned and materialized (i.e., extracting data field of interest), then transferred to a hash table in working memory allocated in the main memory.

Near-Data Database Processing The advancement of storage technologies has done the heavy lifting of a large capacity, high-speed database system. While HDDs are still heavily used in a bulk storage system for their cheap cost per bit, SSDs offer much better I/O performance without exorbitant cost increase. They have been increasingly employed in performance-critical systems. It is also worth mentioning that simply swapping HDDs with SSDs does not require any software changes but can provide high performance. Some NVMs further offer a direct access (DAX) mode to their persistent memory, skipping legacy file systems (e.g., Ext4 and XFS). This allows designing software specifically for persistent memory, such as NOVA [187], which is reported to provide much better performance scaling [188]. Several works have explored near-storage computing using embedded processors in SSDs for database operation to efficiently utilize the bandwidth between CPU and storage [109, 110, 112]. Since the internal bandwidth of SSDs can be higher than the external link bandwidth, performing in-place operations such as data filtering can reduce data communication through the narrower external link. However, near-storage programs have to be designed carefully because the embedded processors have limited performance. Also, because the main memory is still orders of magnitude faster, index-based approaches can outperform depending on the selectivity.

To avoid the bottleneck of disk I/O, RDBMS increasingly runs the backend directly on data in memory. However, memory bandwidth is still forming a bottleneck. Near-memory computing exploits large internal memory bandwidth for such data analysis workloads in memory. One of the challenges of near memory processing is the data movement across memory partitions. This is because many operators rely on the scatter/gather operation, so simply offloading computation closer to the memory does not solve the data movement bottleneck. Mondrian Data Engine [35] proposes a near-memory data analytics accelerator using HMC, focusing on the common database operators. It presents an algorithm-hardware co-design to minimize the burden of inter-memory-partition communication. While popular algorithms in RDBMS are designed for better cache utilization (such as index table), near-memory computation prefers simple sequential access patterns. This is because, while the constrained logic capacity can limit the number of in-flight random accesses, near-memory computing can enjoy faster and cheap memory access. For example, sort-merge join is used as an alternative to hash join. Further, they exploit the data permutability (insensitivity of data ordering to the result) to make it unnecessary to reorder the out-of-order arrival messages of inter-node communication.

KVS offers a huge scalable hash table that can be used as a fast cache in the main memory. Memcached is one of the representative KVS systems, and is popular for its simple interface comprised of *get* and *put*. The bottleneck of the KVS system on the main memory is neither in hash computation nor memory access; rather, it is in the network stack. Near-memory computing targets maximizing the amount of DRAM in a system within given system-wide power, area, and network connections, using small in-order cores in the stacked memories to increase data storage efficiency [189]. Minerva [190] proposes near-flash KVS acceleration for MemcacheDB, a persistent version of Memcached. For the *get* operation, the storage processor applies a hash function and traverses the chain of buckets in the storage. For the *put* operation, it applies a hash function and inserts a key-value pair to a bucket. Because the limitation in latency and bandwidth of flash and PCIe links makes host-flash communication expensive, the large available bandwidth in SSD provides performance benefits.

Similar to sparsity in the graph workload, in-memory computing faces several challenges for the database workload. Some of the database operations, such as sort and join, are difficult to support within the existing computation framework of in-memory computing. Likewise, a database generally densifies data while moving it, which reduces the opportunity for data reuse and in-place computing. However, some other operators such as scan can be very efficiently supported, exploiting the parallelism of the memory array using widely studied XNOR operation, etc. Even for these operations, the cost for in-memory computing has to be carefully evaluated with traditional schemes (e.g., index scan) because the performance of the scan algorithm is highly likely to depend on the result selectivity and other factors. Furthermore, it is essential to manifest an optimization schema for a given query plan, as the database workload is generally given as a chain (or a tree) of basic operations.

6.5   GENOMICS

One of the important drivers of the precision medicine revolution has been genome sequencing. Determining the sequence of characters in an individual's genome can enable early diagnosis of rare genetic disorders, determine an individual's predisposition to different diseases and help design better drugs. In this subsection, we discuss some of the computationally intensive steps in genome sequence analysis and the potential for in-/near-memory computation approaches to accelerate some of these steps.

A genome can be considered to be a long string of DNA characters over the alphabet $\{A, C, G, T\}$. The characters are commonly called bases. The human genome has $\sim$6 billion bases across the two DNA strands. The bases A and T, C and G occur in complementary pairs on the two DNA strands. Each A-T or C-G complement combination is called a *basepair (bp)*. Genome sequencers read the DNA by breaking it into millions or billions of short 100–10K character fragments called *reads*. These reads are assembled together in a computationally intensive read alignment step that determines the best location of each read by matching against a reference genome. To determine the best location for each read in a reference genome, read aligners commonly use the *seed-and-extend* heuristic. First, the seeding step identifies promising locations to map the read, based on the locations of short 15–30 character exact matching substrings (called seeds) between the read and the reference. Second, the seed-extension step scores each of the candidate locations from seeding to determine the best scoring location for the read. To reduce work done during seed-extension, read aligners either group nearby seeds in the reference in a *chaining* step and score them together, or use *seed filtration* heuristics to avoid scoring regions in the genome with very few seeds from the read. Since each read can be aligned independently, read alignment computation is embarrassingly parallel.

With the advent of second- and third-generation genome sequencers, there has been a dramatic reduction in the cost of genome sequencing and corresponding increase in sequencing machine throughput. For instance, in the last decade, the cost to sequence a human genome has dropped from $10 million to less than $1000 today. The reduction in cost and increase in sequencing throughput has led to an exponential growth of sequencing data, at a rate that far outpaces Moore's law (doubling every 12 months). For example, the Whole Genome Shotgun (WGS) project in GenBank has 9.2 trillion bases as of October 2020.

The deluge of genomic data poses an increasing computational burden on sequencing analysis software pipelines running on commodity CPUs. These pipelines routinely process gigabyte to terabyte-scale datasets and employ several data-intensive computational kernels for exact and/or approximate string matching. Many of these kernels perform simple computational operations (e.g., string comparison and integer addition) and spend significant portions of their runtime in data-movement from memory to computational units. These kernels also perform fine-grained, irregular memory lookups into gigabyte-size index structures (e.g., hash tables) and have little spatial or temporal locality making them memory-bandwidth limited on commodity CPUs. It is also common to observe sequential dependent memory lookups in these algorithms (e.g., pointer chasing), making them memory-latency limited. Below we provide an overview of some of these data-intensive kernels and also discuss near-data acceleration approaches for these kernels proposed in literature.

6.5.1   DATA-INTENSIVE KERNELS IN GENOMICS

Figure 6.6 shows some common data-intensive genomics kernels that can benefit from in-/near-memory processing.

Exact String Matching: One of the most common applications of exact string matching in genomics is the seeding step of read alignment. Seeding is done by matching up to k-characters (k-mer) from the read in an index of the reference. Hash tables and FM-index are the most common indexing data structures used for seeding. A hash table can support the lookup of k-characters from the read in a single iteration but requires more memory space ($\sim$30 GB for the human genome). In contrast, the FM index is compact ($\sim$4.3 GB for the human genome) and can efficiently identify variable length seeds in the read by performing single-character lookups. Figure 6.6a shows the FM-index constructed for a sample reference (R). The FM-index consists of: (1) the *suffix array (SA)*, which maps sorted suffixes to their start positions in the reference genome R; (2) the *Burrows Wheeler Transform (BWT)* array, defined as the last column of the matrix of sorted suffixes of the reference; (3) the *count table (Count)* which stores the number of characters in R lexicographically smaller than a given character c (sort order: $\$ < A < C < G < T$, where $\$$ is the sentinel character denoting the end of the string); and (4) the *occurrence table (OCC)* which stores the prefix-sum of occurrences of each character up to a given index in the *BWT* array:

$$start_{i+1} = Count(c) + OCC(start_i - 1, c)$$
$$end_{i+1} = Count(c) + OCC(end_i, c).$$

$$(6.1)$$

The FM-index supports the backward search of a read query Q of length $|Q|$ in $\mathcal{O}(|Q|)$ iterations. Equation (6.1) shows the computation performed in each iteration of backward search. *start* and *end* pointers indicate the range in the *OCC* table containing the matching substring from the read. The computation resembles pointer chasing with two memory lookups to the gigabyte-sized *OCC* table per query character and simple integer addition between iterations. Furthermore, these memory lookups are irregular with little spatial or temporal locality as can

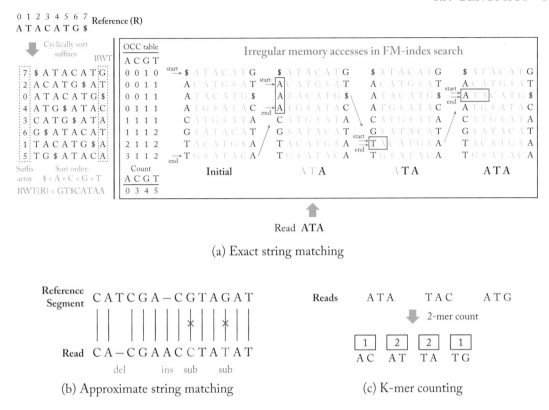

Read **ATA**

(a) Exact string matching

Reference
Segment   C A T C G A − C G T A G A T

Read   C A − C G A A C C T A T A T
           del   ins   sub   sub

(b) Approximate string matching

Reads   A T A   T A C   A T G

2-mer count

1	2	2	1
AC	AT	TA	TG

(c) K-mer counting

Figure 6.6: Common computation kernels in genomics that can benefit from near-data processing. (a) Exact String Matching, (b) Approximate String Matching, and (c) K-mer counting.

be seen from the dotted arrows in Figure 6.6a, making FM-index search memory bandwidth bound. In practical implementations, the *OCC* table is downsampled to reduce index size. Missing *OCC* table entries can be computed from sampled entries using additional symbol comparison and counting operations (e.g., count of "A" in $BWT(4, 6]$, as shown in Figure 6.7).

Approximate String Matching: The seed-extension step of read alignment extensively uses approximate string matching. Starting from either end of the seed, the read is scored against each of the candidate reference locations using a dynamic-programming algorithm for approximate string matching like Smith–Waterman or Needleman–Wunsch. The affine-gap [191] scoring function is commonly used for scoring pairwise DNA alignments. It weighs different edit types (substitutions, insertions, deletions) differently using three matrices H, E, and F as shown in Equation (6.2). To mimic biological evolutionary processes, a linear cost function for scoring

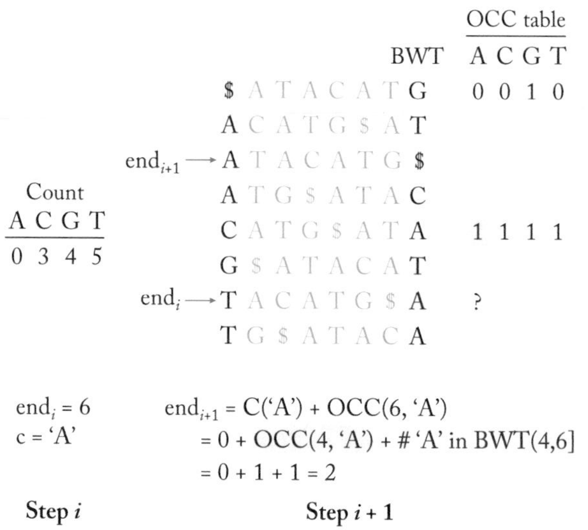

OCC table

		BWT	A C G T
	$ A T A C A T G	0 0 1 0
	A C A T G $ A T	
$end_{i+1} \rightarrow$	A T A C A T G $	
Count	A T G $ A T A C	
A C G T	C A T G $ A T A	1 1 1 1
0 3 4 5	G $ A T A C A T	
$end_i \rightarrow$	T A C A T G $ A	?
	T G $ A T A C A	

$end_i = 6$ $end_{i+1} = C(\text{‘A’}) + OCC(6, \text{‘A’})$

$c = \text{‘A’}$ $= 0 + OCC(4, \text{‘A’}) + \# \text{‘A’ in BWT}(4,6]$

                $= 0 + 1 + 1 = 2$

Step i   **Step $i + 1$**

Figure 6.7: FM-index search with downsampled OCC table.

insertions and deletions is used (cost $= q + l * e$, where l–insertion/deletion length and $q > e$):

$$H_{ij} = \max\{H_{i-1,j-1} + s(i, j), E_{ij}, F_{ij}\}$$
$$E_{i+1,j} = \max\{H_{ij} - q, E_{ij}\} - e \qquad (6.2)$$
$$F_{i,j+1} = \max\{H_{ij} - q, F_{ij}\} - e.$$

For a pair of query Q and reference R strings, H_{ij} is the similarity score for substrings $Q[0, i]$ and $R[0, j]$ and $s(i, j)$ is a pre-computed similarity score between characters $Q[i]$ and $R[j]$. Equation (6.2) has time and space complexity $\mathcal{O}(|Q||R|)$.

K-mer Counting: Another common application in genomics is k-mer counting. A k-mer is a short substring of length k from the read. K-mer frequency estimation from sequenced reads is one of the most common tasks in bioinformatics. For instance, k-mer frequency histograms are used to filter out sections of reads containing low-frequency k-mers (that are likely to have resulted from erroneous sequencing [192]) and high-frequency k-mers (which can confound genome assembly [193]). Bloom filters and hash tables are the most common data-structures used for k-mer counting. The Bloom filter is used to identify unique k-mers in sequenced reads, while the counts for other non-unique k-mers are stored in a hash table (Figure 6.6c). Common values of k are between 25 and 55. K-mer counting involves several random accesses to large hash tables especially for reads from genomes containing repetitive k-mer sequences like humans. It is dominated by LLC misses and data movement from main memory with little computation.

6.5.2   NEAR-DATA PROCESSING ARCHITECTURES FOR GENOMICS

Several efforts have identified the high data movement overheads in genomics kernels and proposed near-memory computing approaches to accelerate them. The high internal bandwidth and available parallelism in conventional LRDIMMs can be used to accelerate memory-bandwith limited FM-index search. For instance, these LRDIMMs can be augmented with simple processing elements to accelerate the symbol comparison and integer addition operations performed in FM-index search [194], without involving the CPU for these operations. Conventional ReRAM chips can also be re-purposed as Hamming distance compute units and augmented with lookup table based adders to accelerate the symbol comparison and count operations in FM-index search [195].

Other works have accelerated the seed filtration step by leveraging the bit-parallelism of DRAM [196], SRAM [197], and ReRAM [198] memory arrays. GRIM-Filter [196] partitions the reference genome into several bins. Each bin is represented as a 4^k-bit bitvector with each bit in the bitvector indicating the presence/absence of the corresponding k-mer in the bin. Each bin is mapped to a column of DRAM in 3D-stacked memories. Similar to the automata processor, the row address is used to stream input seeds from the read. A simple DRAM read enables massively parallel comparison of seeds to bins and returns all the bins containing a particular seed in parallel. Lightweight comparison and accumulation units embedded in the logic layer of these 3D-stacked memories filter out bins with very few seeds from the read. In a similar spirit, GenCache [197] is a recent work that has also accelerated seed filtration by augmenting the large on-chip SRAM scratchpads of genome sequence alignment accelerators like GenAx [199] and Darwin [200] with bitline computational logic to support parallel bitwise operations.

For genomics kernels that perform only simple computation, resistive memories find utility mostly as similarity search engines instead of matrix-vector multiplication engines. They have been used to accelerate variants of the dynamic programming-based approximate string matching algorithm. PRinS [201] and BioSEAL [202] implement the dynamic programming algorithm using associative processing on resistive CAMs (ReCAMs). The massively parallel match/mismatch functionality provided by CAMs makes them good candidates for sequence search applications. In associative processing any compute operation is implemented by comparing each entry of its truth table with the input operands. However, associative processing comes with high write overheads and requires significant intermediate storage. Instead of associative processing, RAPID [203] convert the dynamic programming algorithm as a sequence of XOR, addition, minimum/maximum operations and extend conventional ReRAMs to implement them efficiently. They also propose a hierarchical compute-enabled H-tree architecture to reduce write operations and data movement overheads. RADAR [204] accelerates the simpler ungapped version of approximate string matching (no insertions/deletions) used in read-aligners like BLASTN. They observe that ungapped matching does not require writes and leverage the row-level and array-level parallelism of 3D ReCAMs to find all matches of the read against the reference sequence.

K-mer counting has been successfully accelerated using high-bandwidth memories augmented with simple processing elements to perform hash function computation [205, 206]. NEST [207], a recent DIMM-based k-mer counting accelerator, proposes to augment conventional LRDIMMs with efficient intra-DIMM communication mechanisms to improve bandwidth utilization and reduce load imbalance in k-mer counting.

In addition to ReRAM, bitline computing operations (XNOR, addition, and majority) in SOT-MRAMs have also been leveraged to accelerate read alignment [208, 209] and other genome analysis steps like de-novo assembly [210] and base calling [211].

Pitfall Why or why not Automata Processor (AP) for genomics?

- The AP has been successful in accelerating several pattern matching applications, but a few things need to be kept in mind when trying to map genome sequence analysis applications to the AP.

- The AP is advantageous when matching an input stream against a database of several thousand known motifs. It has been used to accelerate NP-hard search problems in bioinformatics, e.g., (l, d) motif-search [167], which involves searching for all patterns of length l in the input stream differing from the database of known motifs by atmost d substitutions.

- The AP is less suited for other genome sequence analysis applications like read alignment. The Levenshtein Automata which computes the edit distance for approximate string matching can be mapped to the AP. However, the Levenshtein automata is string-dependent. To support the matching of input reads against billions of patterns in the human genome, the AP will need to be frequently reconfigured, leading to performance overheads.

- Furthermore, the simple edit distance scoring metric to measure string similarity is inadequate for applications like read alignment. It is often replaced with more biologically relevant scoring schemes like the affine gap metric that weighs different types of edits (substitutions, short/long insertions and deletions) differently. It is non-trivial to extend the Levenshtein automata to support affine gap scoring and map it onto the AP.

CHAPTER 7

Programming Models

In previous chapters, we discussed several in-/near-memory computing accelerators for several applications in machine learning, databases, graph processing, and genomics. In this chapter we discuss the programming interfaces exposed by these accelerators and the trade-offs involved. We begin by discussing some of the programming models proposed for the early near-memory processing architectures in Section 7.1. Later, we discuss the design trade-offs made in recent domain-specific programming models in Section 7.2. We end by explaining the data-parallel programming models and compilation strategies adopted by recent in-memory computing architectures in Section 7.3.

7.1   HISTORICAL EXECUTION MODELS

To take advantage of intelligent memory systems, many historical execution models for near-memory processing architectures explore several techniques ranging from encoding simple computation in typical memory requests and network messages to binding full functions in virtual memory pages. Active Pages [212] is one such novel computation model that aims to offload the data manipulation parts of applications to intelligent memory systems and keep the processor busy in executing complex operations (e.g., floating point multiply). Each active page consists of data (e.g., array data structure) and a set of associated functions to operate on the data (e.g., find, insert, delete). While active pages associate a set of functions with a page of data (512 KB in the original proposal), the DIVA [213] near-memory processing architecture makes use of small messages called *parcels* [214] to communicate data across DIMM modules and perform coordination and synchronization. A parcel resembles a small active message [215] (header + 256 bit payload in the original proposal) that includes the virtual address of the data to be modified by the parcel and an associated operation on the data. Similar to parcels, Kogge proposes the *traveling threadlets* concept [216]. In this proposal, passive memory requests are transformed into active threadlets that can migrate from physical memory chip to another. A threadlet includes a short program (that performs a very specific operation in memory, e.g., simple synchronization primitives like test-and-set, compare-and-swap that are long-latency operations on CPUs), a set of working registers and program state information (e.g., active, suspended).

7.2 RECENT DOMAIN-SPECIFIC MODELS

In this section, we discuss the programming models explored in recent domain-specific in-/near-memory processing architectures. To enable ease of adoption, many of these approaches integrate their accelerators into existing high-level domain-specific programming frameworks. These frameworks lower the barrier of entry for programmers who can now focus on application-specific optimizations, without worrying about issues like application to hardware mapping, data management and scheduling complexity. For instance, machine learning and graph processing are both framework-based. In-/near-memory machine learning accelerators like [146] can be programmed using popular high-level frameworks like TensorFlow, Microsoft CNTK, PyTorch, and Caffe2. Programs written in these domain-specific languages are first compiled to intermediate computation graphs and later optimized based on the underlying hardware. ONNX (Open Neural Network Exchange Format) is a popular format for representing these computation graphs with support for multiple programming frameworks and multiple hardware. As a result, many in-/near-memory machine learning approaches extend the ONNX front-end and back-end interfaces to support execution on their own accelerators, without exposing these accelerator-specific interfaces to the programmer. In a similar memory manner to machine learning, lightweight extensions to graph processing frameworks like GraphLab, GraphX, IBM System G, and GraphMat have also been proposed to support in- or near-memory accelerators for graph processing [182].

7.2.1 KEY DESIGN DECISIONS

Several issues need to be considered when designing a programming model for in- and near-memory processing architectures [217–219]. It must be expressive enough to make best use of the underlying hardware, while at the same time aim to reduce programmer burden.

Offload Granularity   Prior in-/near-memory acceleration approaches have proposed offloading either a `single-operation`, `bulk operations`, `functions within an application` or `entire applications` to intelligent memory architectures [220]. Significant benefits from instruction-level offloading have been observed for graph processing. ISA extensions have been proposed for each offloaded operation in some near-memory computing architectures with fixed-function units [41]. Since instructions are translated on the host, they need not be additionally decoded in the memory controller prior to offload [221]. In GraphPIM [182], each atomic instruction to a graph vertex (allocated in designated uncacheable regions), is offloaded as a separate instruction. Seshadri et al. [49] and Aga et al. [10] propose offloading bulk bitwise operations to compute-enabled DRAM and SRAM respectively and demonstrate significant benefits for several data analytics kernels. Other approaches use compiler pragmas or library calls to designate functions for offloading to memory. For example, 98.1% speedup was observed for GEMM kernels [222] by offloading their packing and quantization functions. While there are a few systems that offload complete applications to memory (e.g., HMC-based Tesseract [37] for graph processing), it is rare to offload entire applications because of the complexity involved.

Programming Complexity   As discussed earlier, many of the applications that benefit from near-memory processing are framework-based. Many data analytics workloads shard and process data in multiple servers in parallel and are written in distributed computing frameworks like Spark, Hadoop, and MPI. To ease programmer burden, many near-memory processing architectures also adopt a similar approach and extend these high-level programming frameworks and customize them for their own architectures. NDA [34], NDC [36], Tesseract [37], and Active Memory Cube [223] are examples of in-/near-memory computing architectures with a MapReduce/MPI based programming model. Building on these frameworks, Alian et al. [224] propose a TCP/IP-based interface between the host and near-memory compute capable servers to enable programmer-transparent distributed computing across multiple server nodes.

To provide greater flexibility and support multiple high-level frameworks, other in- and near-memory processing approaches [44, 155] base their programming models on lower-level programming models like OpenCL, widely used in heterogeneous computing. However, OpenCL increases programmer burden. At the other end of the spectrum, to provide maximum flexibility and fine-grained control, many others also propose custom programming models with low-level APIs and ISA extensions. However, these are more well suited for domain experts with detailed knowledge of both program behavior and the memory architecture.

There have also been efforts to automate the identification of offload candidates in high-level programs. For example, static analysis on CUDA programs has been used to tag profitable basic blocks [225] and optimize the data mapping and scheduling [226] of these blocks on in-/near-memory architectures.

Address Translation and Cache Coherence   Support for virtual-to-physical address translation is another important requirement for computation kernels involving pointer-based data structures like trees and linked lists that are common acceleration candidates for in-/near-memory processing architectures. Although some approaches embed address translation units within intelligent memories, these increase system complexity. Recent in-/near-memory accelerators favor simpler host-based translation mechanisms by restricting the regions of physical memory operated on by the accelerator.

In addition to virtual address translation, providing hardware support for a coherent shared memory abstraction to the programmer (unifying both the host and near-data accelerator memories) can greatly simplify programming. However, fine-grained coherence protocols like MESI can greatly increase host-memory network traffic. Recent work CoNDA [42], observes that although some near-data accelerated kernels exhibit significant data sharing between the host and the accelerator, there are rarely cache line conflicts with the host. CoNDA optimistically executes the kernel, assuming all permissions are granted and tracks modified cache lines. These modified cache line addresses are compressed and transferred to the host for conflict detection. If conflicts are detected, the kernel is re-executed.

7.3   DATA PARALLEL PROGRAMMING MODELS

With the increased generality of operators supported in in-memory computing devices, efforts have been made to expose the benefits of such devices to a broader context of applications. Although in-memory computing can offer massive computation horsepower as we have witnessed, it is not always straightforward for an application to take advantage of it. Likewise, since one source of performance gain is from reduced data movement, proper handling of the input data takes on a significant aspect of execution flow design. Programming models need to be aware of in-memory devices' constraints, but at the same time, they must offer a programmer-friendly interface and transparency of the data flow. The compilers then need to explore data mapping and operation scheduling to fully expose the parallelism in both the input operation sequence and the hardware. In this section, we show the opportunities and challenges of general-purpose in-memory computing and programming interfaces.

Opportunities   Despite the significant performance gain of in-memory computing devices, most accelerator works have relied on manual mapping of computation kernels to the memory arrays or through domain-specific languages (DSLs) designed specifically for a class of workload, making it difficult to configure it for diverse applications. One way to combat this problem is to design a programmable in-memory processor architecture that works with a generally used programming framework. It has been reported that general-purpose in-memory processors have the potential to improve the performance of data-parallel application kernels by an order of magnitude or more.

The efficiency of an in-memory processor comes from two sources. The first is massive data parallelism. For example, NVMs are composed of several thousands of arrays. Each of these arrays can be transformed into a single instruction multiple data (SIMD) processing unit that can compute concurrently. The second source is a reduction in data movement, by avoiding shuffling of data between memory and processor cores. On the other hand, NVMs have relatively large read/write latency, leading to large execution latency for arithmetic operations. An expressive programming model for in-memory computing exposes these unique characteristics and opportunities of computable memory devices to applications and programmers without significant efforts to rewrite the programs.

Challenges   One of the challenges for general-purpose in-memory computing is to have a rich set of computation primitives to address a broad set of data-intensive applications. While many works support logical and fixed-point arithmetic operations, many data-parallel workloads require complex arithmetic operations such as transcendental functions and high precision arithmetic such as floating points. Manipulation of mantissa based on exponents in an in-memory vector architecture is a non-trivial challenge. Another challenge for in-memory computing is designing the interface between the CPU core and compute memories, execution model, and memory addressing. Operands of in-memory operations need to be aligned on a bitline ALU, constraining them to specific locations in memory.

The programming model creates a direct and significant impact on programmability and architecture design. While simple models (e.g., wide SIMD [11]) simplify the hardware, it comes at a cost of limited flexibility. On the other hand, guaranteeing too much freedom may result in over-provisioning of hardware resources in order to handle all computation and communication patterns. From the programmers' perspective, it is desirable to follow or make compatible with the widely used programming models to reduce the efforts to rewrite programs and offer cross-platform compatibility. Existing programming models for GPU and machine learning already provide great expressibility for data-parallel programs which are the main target of in-memory computing. The following section will introduce two works that use TensorFlow, Google's machine learning framework, for SIMD processing in ReRAM, and GPU programming frontends (CUDA/OpenACC) for SIMT-like processing in SRAM. These approaches are coupled with compiler optimizations for resource allocation, instruction scheduling, and parallelization to better exploit the computation resources in memory.

7.3.1   SIMD PROGRAMMING

IMP [227] views ReRAM based in-memory computation devices as a large SIMD processor, and proposes to utilize the underlying parallelism in the hardware by merging the concepts of data-flow and SIMD vector processing. Data-flow explicitly exposes the Instruction Level Parallelism (ILP) in the program, while vector processing exposes the Data Level Parallelism (DLP). They use Google's TensorFlow [228], a popular programming model for machine learning, based on their observation that TensorFlow's programming semantics is a good marriage of data-flow and vector-processing that can be applied to more general applications.

TensorFlow (TF) offers a suitable programming paradigm for data-parallel in-memory computing. First, nodes in TF's data flow graphs (DFGs) can operate on multi-dimensional matrices. This feature embeds the SIMD programming model and facilitates exposure of Data Level Parallelism (DLP) to the compiler. Second, irregular memory accesses are restricted by not allowing subscript notation. This feature benefits both programmers and compilers. Programmers do not have to convert high-level operations (e.g., vector addition) into low-level procedural representations (e.g., for-loops with indices for memory access). The compiler can easily understand the operations and the memory access pattern. Third, the DFG naturally exposes Instruction Level Parallelism (ILP). This can be directly used by a compiler for Very Long Instruction Word (VLIW) style scheduling to further utilize the underlying parallelism in the hardware without implementing complex out-of-order execution support. Finally, TensorFlow supports a persistent memory context in nodes of the DFG. This is useful in a situation where data can be reused across different kernel invocations. TensorFlow supports simple control flow by a select instruction. A select instruction takes three operands and generates output $O[i] = Cond[i] ? A[i] : B[i]$, and is converted into multiple selective move instructions.

Compiling and Executing SIMD Programs   The TensorFlow (TF) programs are essentially DFGs where each operator node can have multi-dimensional vectors, or tensors, as operands. A DFG that operates on one element of a vector is referred to as a module by the compiler. The IMP compiler transforms the input DFG into a collection of data-parallel modules with identical machine code. At runtime, a code module is instantiated many times and processes independent data elements. The compiler also performs several DFG optimizations such as unrolling high-dimensional tensors, merging DFG nodes to utilize n-ary ReRAM operations, pipelining compute and write-backs, maximizing ILP within a module using VLIW style scheduling, and minimizing communication between arrays.

At runtime, different instances of a module execute the same instructions on different elements of input vectors in a lock-step manner. A module is generated by unrolling a single dimension of multi-dimensional input vectors as shown in Figure 7.1. At kernel launch time, module instances are dynamically created in accordance with the input vector length. Each module is composed of one or more Instruction Blocks (IB), as shown in Figure 7.1. An IB consists of a list of instructions that will be executed sequentially and is responsible for executing a group of nodes in the DFG. Multiple IBs in a module may execute in parallel to expose ILP.

Rows in the ReRAM array are viewed as a SIMD vector unit with multiple SIMD lanes. Each IB is mapped to a single lane. To ensure full utilization of all SIMD lanes in the array, the runtime maps identical IBs from different instances of the same module to an individual array, as shown in the last row of Figure 7.1. This mapping results in correct execution because all instances of a module have the same set of IBs. Furthermore, IBs of a module are greedily assigned to nearby arrays to minimize the communication latency between IBs.

7.3.2   SIMT PROGRAMMING ON SRAM

TensorFlow provides a suitable programming paradigm for parallel in-memory computing in main memory by the explicit notation of vector datatype, which can be augmented to SIMD computing, and the restricted data-flow, which facilitates architecture design exploration. However, not all programs are vectorizable, and some applications with irregular memory access may face difficulties in these frameworks. Moreover, in-memory computing in main memory has a relatively strict restriction on data flow and data structure, since internal interconnect bandwidth is not designed for arbitrary communication patterns, which can also impact non-vectorizable programs.

Duality Cache (DC) [63] adopts CUDA/OpenACC as a programming model and proposes a compiler which can translate arbitrary CUDA/OpenACC programs to the DC's ISA for in-cache computing on SRAM. DC architecture extends the compute capability of Compute Caches [10] and Neural Cache [11] for general-purpose computing. The compiler allocates

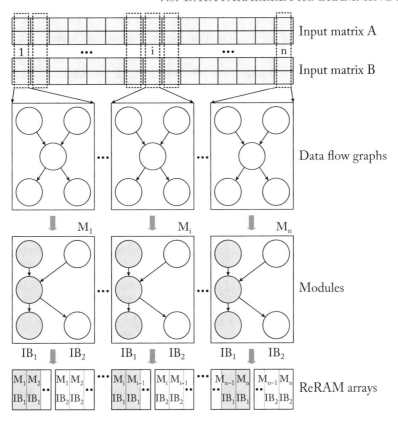

Figure 7.1: Execution model of IMP [227].

resources, schedules VLIW instructions, and conducts several optimizations exploiting unique opportunities in our in-cache architecture.

CUDA describes kernels as multi-threaded programs and groups threads into warps. In a warp, threads are executed in a synchronized manner. Inter-thread synchronization and sharing are allowed within a group of threads called thread block or Cooperative Thread Array (CTA). In other words, different CTAs are independent and can be scheduled and executed in any order. In-cache computing benefits from this programming model from two aspects. First, CUDA is a popular and widely used framework across different fields spanning from scientific computing to machine learning. Leveraging it for an in-cache computing architecture with direct translation or with trivial source code changes will achieve portability and opportunity to use the existing software. Second, having independent CTAs entails minimum network resources for inter-thread communications that happen locally within a CTA.

DC also supports OpenACC. OpenACC provides OpenMP-like pragmas to programmers, making it easier to convert existing serial programs to parallel programs. Currently, commodity OpenACC compilers support multi-thread, multi-core CPUs, and NVIDIA GPUs. OpenACC is characterized by its ability to describe fine-grained interleaving of serial computation on the host and parallel kernels to be executed on an accelerator (e.g., GPU) using pragma. While GPUs tend to face a communication bottleneck for those OpenACC programs with frequent host-device communication, DC enables seamless execution between host code and kernel code, as caches share the same memory hierarchy as the host.

Executing and Compiling SIMT Programs   The SIMT execution model of DC is simpler and coarser-grained when compared to GPUs. DC allows caches to act in two modes: accelerator mode and cache mode. In the accelerator mode, a bitline in a cache array becomes one thread lane of a SIMT processor. In the in-cache computing mode, the registers and compute units are identical. DC assigns registers in a thread to the bitline and performs the computation in place. Operands are vertically aligned within the registers mapped on the bitline. In the cache mode, the cache arrays are part of the processor's traditional multi-level memory hierarchy.

DC dedicates an entire bank with four SRAM arrays to a Thread Block (TB) to provide a sufficient number of registers per thread and prevent frequent register spilling. One SRAM array has 256-bit cells along a bitline, thus can afford only eight 32-bit bit-serial registers as shown in Figure 7.2b. By allocating 256 threads to a bank of 4 SRAM arrays, DC can afford thirty-two 32-bit bit-serial registers per thread. Thus, each thread in a TB is virtually mapped to multiple arrays in a bank, and each member array has a slice of registers. Inter-thread communication is only allowed within TB. This design choice is made to balance programmability and hardware complexity. A 256×256 local crossbar in the C-Box is utilized to shuffle/broadcast CTA local data as shown in Figure 7.2a.

Mapping a TB to a bank of four arrays enables each array to execute a different operation in the same cycle. DC performs VLIW-like instruction scheduling. VLIW allows DC to exploit ILP in the program with low hardware complexity, since the compiler handles complicated ILP aware scheduling. DC performs the computation in a bit-serial manner. Each bitline acts as a computation unit, and all bitlines in an array perform the same operation as in a SIMD processor.

The DC compiler translates CUDA code to a VLIW SIMD ISA. Unlike traditional VLIW architectures, the DC architecture has to take operand locality into account; all operands need to reside in the array where the operation is executed, otherwise, the operands have to be explicitly copied.

Register pressure and efficient VLIW instruction scheduling are inseparable problems. In DC, instruction scheduling is tightly coupled with resource allocation. While many compilers for VLIW architecture schedule instructions first before register allocation to maximize parallelism utilizing abundant register resources shared by many execution units, DC has a limited number of private registers, which may result in frequent register spilling. On the other hand, resource-allocation-first approaches often introduce many false dependencies in return for minimized register usage, which can reduce available parallelism. DC tackles this problem by performing resource allocation and instruction scheduling at the same time. DC uses Bottom-Up

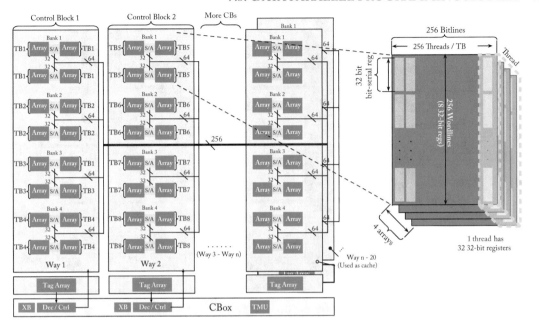

Figure 7.2: In-Cache SIMT execution model and architecture of DC [63].

Greedy (BUG) [229] as the baseline scheduling algorithm, and linear scan register allocation as the baseline resource allocation algorithm. By taking register pressure into account while performing instruction scheduling, the compiler can pick a better strategy to balance parallelism and register spilling.

Pitfall In-memory computing can be easily integrated into the existing software/hardware stack.

This is not true because of the following reasons.

- In-memory computing often requires special data mapping and addressing. For example, in-memory vector computing of DRAM and NVMs usually requires operands to be aligned in the same columns, and bit-serial computing further requires the operand bits to be transposed.

- To allow in-memory computing to use a portion of memory, the system has to support a flexible allocation of the region for a scratchpad-like use. For example, several works [230, 231] have re-purposed SRAM caches for a local scratchpad. These works generally require small changes to the OS and hardware. DRAM generally uses XOR-based address mappings [232] to exploit bank and channel parallelism by distributing consecutive cache blocks in the physical address space across DRAM banks and channels. In order to align a page to the same sub-array, the OS has to allocate a physical page of a specific address, and the data should be written to specific cache blocks, in accordance with the XOR address map of the CPU's microarchitecture.

• In addition, efficient on-demand data transformation (e.g., transposition) may need specialized hardware. For example, Neural Cache [11] uses Transpose Memory Unit (TMU), which uses 8-T transpose bit-cells to transpose bits. The memory controller may need to manage operand addressing and data load/store for in-memory computing to avoid costly communication with the host CPU.

• In-memory computing may need to support ECC-enabled high reliability and address potential security issues.

CHAPTER 8

Closing Thoughts

We discussed the foundations of compute capable memories that span diverse memory technologies (SRAM, DRAM, non-volatile memories), capable of accelerating several data-intensive applications (cognitive computing, machine learning, data analytics, graphs, genomics).

Our narrative, cumulative of extensive research in this field, showcases the significant potential of this technology as a candidate for extending Moore's law beyond specialization of compute. As the general-purpose core's efficiency flat-lined over the past decade, both industry and academia have wholeheartedly embraced customization of computational units. It is high time for us to think about customizing memory units as well. There are many ways that one could think of customizing memory, but turning it into powerful accelerators is one of the more exciting avenues to pursue.

While there is a significant body of research on individual layers of memory, and specific memory technologies, a unified framework with custom memory hierarchy solutions has been lacking. For example, which level in the memory hierarchy is most profitable for conducting in-memory or near-memory operations? Figure 8.1 shows the relative energy per access, delay, and metrics for calculating available compute parallelism of different memory technologies. The parallelism can be estimated based on available sensing amplifiers (bitline peripherals) per unit area. Available parallelism is also dependant both on memory bit-cell type and how the memory is organized (e.g., cache, main memory, or storage DIMMs). For example, while NAND-Flash and DRAM have a small cell size, their available parallelism can be low because a large number of cells in an array share the same set of sensing amplifiers (low SA density). In-/near-memory computation characteristics will be similar to a read access or a read access followed by a write access, based on the operation/ISA.

Computing with NVMs has different trade-offs compared to computing with SRAM or DRAM. Since NVMs are more stable against data corruption, they can support operations across multiple wordlines. Due to their high density, NVMs can accommodate large datasets which dwarf SRAMs. Higher density also increases data level parallelism of in-place computation. On the other hand, memory computation in NVMs (STT-RAM and ReRAM) can be 1–2 orders of magnitude slower and requires significantly higher energy per bit when compared to SRAMs. Further, NVMs have limited endurance (and high write energy/delay) which curtails the number of writes the memories can reliably sustain. Similarly, DRAMs pose their own unique challenges. Given the wide spectrum of memory technologies and their differentiated

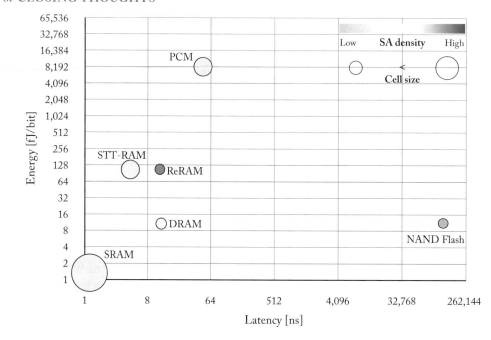

Figure 8.1: Energy, latency, and parallelism characteristics of various memory technologies.

compute capabilities, customizing the memory hierarchy for specific application domains may yield significant benefits and is an interesting future work.

As a closing thought, we hope the large body of work in this field will spur on industry adoption and prototypes. Until recently, we have viewed computing and memory units as two separate entities. Even within a processor, caches and computational logic have operated as two separate entities that served different roles. The time has come to dissolve the line that separates them.

Bibliography

[1] Reetuparna Das. Blurring the lines between memory and computation. *IEEE Micro*, 37(6):13–15, 2017. DOI: 10.1109/mm.2017.4241340 1

[2] Wm. A. Wulf and Sally A. McKee. Hitting the memory wall: Implications of the obvious. *SIGARCH Comput. Archit. News*, 23(1):20–24, March 1995. DOI: 10.1145/216585.216588 1

[3] Maya Gokhale, Bill Holmes, and Ken Iobst. Processing in memory: the Terasys massively parallel PIM array. *Computer*, 1995. DOI: 10.1109/2.375174 1, 44

[4] Yi Kang, Wei Huang, Seung-Moon Yoo, D. Keen, Zhenzhou Ge, V. Lam, P. Pattnaik, and J. Torrellas. FlexRAM: Toward an advanced intelligent memory system. In *Computer Design, (ICCD'99) International Conference on*, 1999. DOI: 10.1109/iccd.1999.808425 1

[5] Peter M. Kogge. Execube-a new architecture for scaleable MPPs. In *Parallel Processing, ICPP. International Conference on*, vol. 1, 1994. DOI: 10.1109/icpp.1994.108 1

[6] Mark Oskin, Frederic T. Chong, and Timothy Sherwood. Active pages: A computation model for intelligent memory. In *Computer Architecture, Proceedings. The 25th Annual International Symposium on*, 1998. DOI:10.1109/isca.1998.694774 1

[7] David Patterson, Thomas Anderson, Neal Cardwell, Richard Fromm, Kimberly Keeton, Christoforos Kozyrakis, Randi Thomas, and Katherine Yelick. A case for intelligent RAM. *Micro, IEEE*, 1997. DOI: 10.1109/40.592312 1

[8] Harold S. Stone. A logic-in-memory computer. *IEEE Transactions on Computers*, 100(1):73–78, 1970. DOI: 10.1109/tc.1970.5008902 1, 15

[9] Paul Dlugosch, Dave Brown, Paul Glendenning, Michael Leventhal, and Harold Noyes. An efficient and scalable semiconductor architecture for parallel automata processing. *IEEE Transactions on Parallel and Distributed Systems*, 25(12):3088–3098, 2014. DOI: 10.1109/tpds.2014.8 2, 69

[10] Shaizeen Aga, Supreet Jeloka, Arun Subramaniyan, Satish Narayanasamy, David Blaauw, and Reetuparna Das. Compute caches. In *IEEE International Symposium on High Performance Computer Architecture (HPCA)*, pages 481–492, 2017. DOI: 10.1109/hpca.2017.21 2, 20, 32, 33, 86, 90

[11] Charles Eckert, Xiaowei Wang, Jingcheng Wang, Arun Subramaniyan, Ravi Iyer, Dennis Sylvester, David Blaaauw, and Reetuparna Das. Neural cache: Bit-serial in-cache acceleration of deep neural networks. In *ACM/IEEE 45th Annual International Symposium on Computer Architecture (ISCA)*, pages 383–396, 2018. DOI: 10.1109/isca.2018.00040 2, 31, 34, 58, 62, 64, 89, 90, 94

[12] Norman P. Jouppi, Cliff Young, Nishant Patil, David Patterson, Gaurav Agrawal, Raminder Bajwa, Sarah Bates, Suresh Bhatia, Nan Boden, Al Borchers, et al. In-datacenter performance analysis of a tensor processing unit. In *Proc. of the 44th Annual International Symposium on Computer Architecture*, pages 1–12, 2017. DOI: 10.1145/3079856.3080246 3

[13] Hoang Anh Du Nguyen, Jintao Yu, Muath Abu Lebdeh, Mottaqiallah Taouil, Said Hamdioui, and Francky Catthoor. A classification of memory-centric computing. *ACM Journal on Emerging Technologies in Computing Systems (JETC)*, 16(2):1–26, 2020. DOI: 10.1145/3365837 5

[14] Fabrice Devaux. The true processing in memory accelerator. In *IEEE Hot Chips 31 Symposium (HCS)*, pages 1–24, Computer Society, 2019. DOI: 10.1109/hotchips.2019.8875680 6, 19

[15] Young-Cheon Kwon, Suk Han Lee, Jaehoon Lee, Sang-Hyuk Kwon, Je Min Ryu, Jong-Pil Son, O. Seongil, Hak-Soo Yu, Haesuk Lee, Soo Young Kim, et al. 25.4 a 20 nm 6 GB function-in-memory DRAM, based on HBM2 with a 1.2TFLOPS programmable computing unit using bank-level parallelism, for machine learning applications. In *IEEE International Solid-State Circuits Conference (ISSCC)*, 64:350–352, 2021. DOI: 10.1109/isscc42613.2021.9365862 6, 19

[16] H. Noyes et al. Microns automata processor architecture: Reconfigurable and massively parallel automata processing. In *Proc. of 5th International Symposium on Highly-Efficient Accelerators and Reconfigurable Technologies*, 2014. 6

[17] Shahar Kvatinsky, Dmitry Belousov, Slavik Liman, Guy Satat, Nimrod Wald, Eby G. Friedman, Avinoam Kolodny, and Uri C. Weiser. Magic—memristor-aided logic. *IEEE Transactions on Circuits and Systems II: Express Briefs*, 61(11):895–899, 2014. DOI: 10.1109/tcsii.2014.2357292 8, 53, 54

[18] Shahar Kvatinsky, Guy Satat, Nimrod Wald, Eby G. Friedman, Avinoam Kolodny, and Uri C. Weiser. Memristor-based material implication (IMPLY) logic: Design principles and methodologies. *IEEE Transactions on Very Large Scale Integration (VLSI) Systems*, 22(10):2054–2066, 2013. DOI: 10.1109/tvlsi.2013.2282132 8, 53

[19] David Patterson, Thomas Anderson, Neal Cardwell, Richard Fromm, Kimberly Keeton, Christoforos Kozyrakis, Randi Thomas, and Katherine Yelick. A case for intelligent RAM. *IEEE Micro*, 17(2):34–44, 1997. DOI: 10.1109/40.592312 14

[20] Andreas Nowatzyk, Fong Pong, and Ashley Saulsbury. Missing the memory wall: The case for processor/memory integration. In *23rd Annual International Symposium on Computer Architecture (ISCA'96)*, pages 90–90, IEEE, 1996. DOI: 10.1145/232973.232984 14

[21] Michael F. Deering, Stephen A. Schlapp, and Michael G. Lavelle. FBRAM: A new form of memory optimized for 3D graphics. In *Proc. of the 21st Annual Conference on Computer Graphics and Interactive Techniques*, pages 167–174, 1994. DOI: 10.1145/192161.192194 15

[22] Duncan G. Elliott, W. Martin Snelgrove, and Michael Stumm. Computational RAM: A memory-SIMD hybrid and its application to DSP. In *Custom Integrated Circuits Conference*, 30:1–30, 1992. DOI: 10.1109/cicc.1992.591879 15

[23] William H. Kautz. Cellular logic-in-memory arrays. *IEEE Transactions on Computers*, 100(8):719–727, 1969. DOI: 10.1109/t-c.1969.222754 15

[24] Chen-Chau Yang and Sik-Sang Yau. A cutpoint cellular associative memory. *IEEE Transactions on Electronic Computers*, (4):522–528, 1966. DOI: 10.1109/pgec.1966.264359 15

[25] William H. Kautz. A cellular threshold array. *IEEE Transactions on Electronic Computers*, (5):680–682, 1967. DOI: 10.1109/pgec.1967.264780 15

[26] Gwangsun Kim, John Kim, Jung Ho Ahn, and Jaeha Kim. Memory-centric system interconnect design with hybrid memory cubes. In *Proc. of the 22nd International Conference on Parallel Architectures and Compilation Techniques*, pages 145–155, IEEE, 2013. DOI: 10.1109/pact.2013.6618812 16

[27] HMC Consortium et al. Hybrid memory cube specification 2.1. hybridmemorycube.org, 2013. 16

[28] JEDEC Standard. High bandwidth memory (HBM) DRAM. *JESD235*, 2013. 16

[29] Berkin Akin, Franz Franchetti, and James C. Hoe. Data reorganization in memory using 3D-stacked DRAM. *ACM SIGARCH Computer Architecture News*, 43(3S):131–143, 2015. DOI: 10.1145/2872887.2750397 17, 18

[30] Joe Jeddeloh and Brent Keeth. Hybrid memory cube new DRAM architecture increases density and performance. In *Symposium on VLSI Technology (VLSIT)*, pages 87–88, IEEE, 2012. DOI: 10.1109/vlsit.2012.6242474 17

[31] Qiuling Zhu, Berkin Akin, H. Ekin Sumbul, Fazle Sadi, James C. Hoe, Larry Pileggi, and Franz Franchetti. A 3D-stacked logic-in-memory accelerator for application-specific data intensive computing. In *3D Systems Integration Conference (3DIC), IEEE International,* 2013. DOI: 10.1109/3dic.2013.6702348 17

[32] Jiayi Huang, Ramprakash Reddy Puli, Pritam Majumder, Sungkeun Kim, Rahul Boyapati, Ki Hwan Yum, and Eun Jung Kim. Active-routing: Compute on the way for near-data processing. In *IEEE International Symposium on High Performance Computer Architecture (HPCA),* pages 674–686, 2019. DOI: 10.1109/hpca.2019.00018 17

[33] Mingyu Gao, Grant Ayers, and Christos Kozyrakis. Practical near-data processing for in-memory analytics frameworks. In *International Conference on Parallel Architecture and Compilation (PACT),* pages 113–124, IEEE, 2015. DOI: 10.1109/pact.2015.22 17, 76

[34] Amin Farmahini-Farahani, Jung Ho Ahn, Katherine Morrow, and Nam Sung Kim. NDA: Near-DRAM acceleration architecture leveraging commodity DRAM devices and standard memory modules. In *IEEE 21st International Symposium on High Performance Computer Architecture (HPCA),* pages 283–295, 2015. DOI: 10.1109/hpca.2015.7056040 18, 87

[35] Mario Paulo Drumond Lages De Oliveira, Alexandros Daglis, Nooshin Mirzadeh, Dmitrii Ustiugov, Javier Picorel Obando, Babak Falsafi, Boris Grot, and Dionisios Pnevmatikatos. The mondrian data engine. In *Proc. of the 44th International Symposium on Computer Architecture,* number CONF, 2017. DOI: 10.1145/3079856.3080233 18, 78

[36] Seth H. Pugsley, Jeffrey Jestes, Huihui Zhang, Rajeev Balasubramonian, Vijayalakshmi Srinivasan, Alper Buyuktosunoglu, Al Davis, and Feifei Li. NDC: Analyzing the impact of 3D-stacked memory+ logic devices on MapReduce workloads. In *IEEE International Symposium on Performance Analysis of Systems and Software (ISPASS),* pages 190–200, 2014. DOI: 10.1109/ispass.2014.6844483 18, 87

[37] Junwhan Ahn, Sungpack Hong, Sungjoo Yoo, Onur Mutlu, and Kiyoung Choi. A scalable processing-in-memory accelerator for parallel graph processing. In *Proc. of the 42nd Annual International Symposium on Computer Architecture,* pages 105–117, 2015. DOI: 10.1145/2749469.2750386 18, 75, 87

[38] Duckhwan Kim, Jaeha Kung, Sek Chai, Sudhakar Yalamanchili, and Saibal Mukhopadhyay. NeuroCube: A programmable digital neuromorphic architecture with high-density 3D memory. *ACM SIGARCH Computer Architecture News,* 44(3):380–392, 2016. DOI: 10.1145/3007787.3001178 18, 62, 65

[39] Mingyu Gao, Jing Pu, Xuan Yang, Mark Horowitz, and Christos Kozyrakis. Tetris: Scalable and efficient neural network acceleration with 3D memory. In *Proc. of the 22nd In-*

ternational Conference on Architectural Support for Programming Languages and Operating Systems, pages 751–764, 2017. DOI: 10.1145/3037697.3037702 18, 62, 65

[40] Mingxuan He, Choungki Song, Ilkon Kim, Chunseok Jeong, Seho Kim, Il Park, Mithuna Thottethodi, and TN Vijaykumar. Newton: A DRAM-maker's accelerator-in-memory (AIM) architecture for machine learning. In *53rd Annual IEEE/ACM International Symposium on Microarchitecture (MICRO)*, pages 372–385, 2020. DOI: 10.1109/micro50266.2020.00040 18

[41] Junwhan Ahn, Sungjoo Yoo, Onur Mutlu, and Kiyoung Choi. Pim-enabled instructions: A low-overhead, locality-aware processing-in-memory architecture. In *ACM/IEEE 42nd Annual International Symposium on Computer Architecture (ISCA)*, pages 336–348, 2015. DOI: 10.1145/2749469.2750385 18, 86

[42] Amirali Boroumand, Saugata Ghose, Minesh Patel, Hasan Hassan, Brandon Lucia, Kevin Hsieh, Krishna T. Malladi, Hongzhong Zheng, and Onur Mutlu. LazyPIM: An efficient cache coherence mechanism for processing-in-memory. *IEEE Computer Architecture Letters*, 16(1):46–50, 2016. DOI: 10.1109/lca.2016.2577557 18, 87

[43] Benjamin Y. Cho, Yongkee Kwon, Sangkug Lym, and Mattan Erez. Near data acceleration with concurrent host access. In *ACM/IEEE 47th Annual International Symposium on Computer Architecture (ISCA)*, pages 818–831, 2020. DOI: 10.1109/isca45697.2020.00072 18

[44] Dongping Zhang, Nuwan Jayasena, Alexander Lyashevsky, Joseph L. Greathouse, Lifan Xu, and Michael Ignatowski. Top-PIM: throughput-oriented programmable processing in memory. In *Proc. of the 23rd International Symposium on High-Performance Parallel and Distributed Computing*, pages 85–98, 2014. DOI: 10.1145/2600212.2600213 18, 87

[45] Mingyu Gao and Christos Kozyrakis. HRL: Efficient and flexible reconfigurable logic for near-data processing. In *IEEE International Symposium on High Performance Computer Architecture (HPCA)*, pages 126–137, 2016. DOI: 10.1109/hpca.2016.7446059 18

[46] Mike O'Connor, Niladrish Chatterjee, Donghyuk Lee, John Wilson, Aditya Agrawal, Stephen W. Keckler, and William J. Dally. Fine-grained DRAM: Energy-efficient DRAM for extreme bandwidth systems. In *50th Annual IEEE/ACM International Symposium on Microarchitecture (MICRO)*, pages 41–54, 2017. DOI: 10.1145/3123939.3124545 20

[47] Tim Finkbeiner, Glen Hush, Troy Larsen, Perry Lea, John Leidel, and Troy Manning. In-memory intelligence. *IEEE Micro*, 37(4):30–38, 2017. DOI: 10.1109/mm.2017.3211117 21

[48] Shuangchen Li, Alvin Oliver Glova, Xing Hu, Peng Gu, Dimin Niu, Krishna T. Malladi, Hongzhong Zheng, Bob Brennan, and Yuan Xie. Scope: A stochastic computing engine for DRAM-based in-situ accelerator. In *MICRO*, pages 696–709, 2018. DOI: 10.1109/micro.2018.00062 21

[49] Vivek Seshadri, Donghyuk Lee, Thomas Mullins, Hasan Hassan, Amirali Boroumand, Jeremie Kim, Michael A. Kozuch, Onur Mutlu, Phillip B. Gibbons, and Todd C. Mowry. Ambit: In-memory accelerator for bulk bitwise operations using commodity DRAM technology. In *50th Annual IEEE/ACM International Symposium on Microarchitecture (MICRO)*, pages 273–287, 2017. DOI: 10.1145/3123939.3124544 21, 22, 23, 24, 86

[50] Vivek Seshadri, Kevin Hsieh, Amirali Boroum, Donghyuk Lee, Michael A. Kozuch, Onur Mutlu, Phillip B. Gibbons, and Todd C. Mowry. Fast bulk bitwise and and or in DRAM. *IEEE Computer Architecture Letters*, 14(2):127–131, 2015. DOI: 10.1109/lca.2015.2434872 21, 24

[51] Brent Keeth, R. Jacob Baker, Brian Johnson, and Feng Lin. *DRAM Circuit Design: Fundamental and High-Speed Topics*, vol. 13, John Wiley & Sons, 2007. 21

[52] Fei Gao, Georgios Tziantzioulis, and David Wentzlaff. ComputedRAM: In-memory compute using off-the-shelf DRAMs. In *Proc. of the 52nd Annual IEEE/ACM International Symposium on Microarchitecture*, pages 100–113, 2019. DOI: 10.1145/3352460.3358260 23, 25

[53] Shuangchen Li, Dimin Niu, Krishna T. Malladi, Hongzhong Zheng, Bob Brennan, and Yuan Xie. DRISA: A DRAM-based reconfigurable in-situ accelerator. In *50th Annual IEEE/ACM International Symposium on Microarchitecture (MICRO)*, pages 288–301, 2017. DOI: 10.1145/3123939.3123977 24, 26, 62, 65

[54] Xianwei Zhang, Youtao Zhang, Bruce R. Childers, and Jun Yang. Restore truncation for performance improvement in future DRAM systems. In *IEEE International Symposium on High Performance Computer Architecture (HPCA)*, pages 543–554, 2016. DOI: 10.1109/hpca.2016.7446093 26

[55] Prashant J. Nair, Dae-Hyun Kim, and Moinuddin K. Qureshi. Archshield: Architectural framework for assisting DRAM scaling by tolerating high error rates. *ACM SIGARCH Computer Architecture News*, 41(3):72–83, 2013. DOI: 10.1145/2508148.2485929 26

[56] Xin Xin, Youtao Zhang, and Jun Yang. ROC: DRAM-based processing with reduced operation cycles. In *Proc. of the 56th Annual Design Automation Conference*, pages 1–6, 2019. DOI: 10.1145/3316781.3317900 26, 27

[57] Neil He Weste and David Money Harris. CMOS VLSI design: A circuits and systems perspective, 2011. 30

[58] Min Huang, Moty Mehalel, Ramesh Arvapalli, and Songnian He. An energy efficient 32-nm 20-mb shared on-die L3 cache for intel® xeon® processor E5 family. *Journal of Solid-State Circuits*, 48(8):1954–1962, 2013. DOI: 10.1109/JSSC.2013.2258815 30

[59] Wei Chen, Szu-Liang Chen, Siufu Chiu, Raghuraman Ganesan, Venkata Lukka, Wei Wing Mar, and Stefan Rusu. A 22 nm 2.5 mb slice on-die l3 cache for the next generation xeon® processor. In *VLSI Technology (VLSIT), Symposium on*, pages C132–C133, IEEE, 2013. 30

[60] Supreet Jeloka, Naveen Bharathwaj Akesh, Dennis Sylvester, and David Blaauw. A 28 nm configurable memory (TCAM/BCAM/SRAM) using push-rule 6T bit cell enabling logic-in-memory. *IEEE Journal of Solid-State Circuits*, 51(4):1009–1021, 2016. DOI: 10.1109/jssc.2016.2515510' 32

[61] Soroosh Khoram, Yue Zha, and Jing Li. An alternative analytical approach to associative processing. *IEEE Computer Architecture Letters*, 17(2):113–116, 2018. DOI: 10.1109/lca.2018.2789424 34

[62] Yue Zha and Jing Li. Hyper-AP: Enhancing associative processing through a full-stack optimization. In *ACM/IEEE 47th Annual International Symposium on Computer Architecture (ISCA)*, pages 846–859, 2020. DOI: 10.1109/isca45697.2020.00074 34

[63] Daichi Fujiki, Scott Mahlke, and Reetuparna Das. Duality cache for data parallel acceleration. In *Proc. of the 46th International Symposium on Computer Architecture*, pages 397–410, 2019. DOI: 10.1145/3307650.3322257 36, 90, 93

[64] Jingcheng Wang, Xiaowei Wang, Charles Eckert, Arun Subramaniyan, Reetuparna Das, David Blaauw, and Dennis Sylvester. 14.2 a compute SRAM with bit-serial integer/floating-point operations for programmable in-memory vector acceleration. In *IEEE International Solid-State Circuits Conference-(ISSCC)*, pages 224–226, 2019. DOI: 10.1109/isscc.2019.8662419 36

[65] Jae-sun Seo, Bernard Brezzo, Yong Liu, Benjamin D. Parker, Steven K. Esser, Robert K. Montoye, Bipin Rajendran, José A. Tierno, Leland Chang, Dharmendra S. Modha, and Daniel J. Friedman. A 45 nm CMOS neuromorphic chip with a scalable architecture for learning in networks of spiking neurons. In *IEEE Custom Integrated Circuits Conference (CICC)*, pages 1–4, 2011. DOI: 10.1109/cicc.2011.6055293 36

[66] Mingu Kang, Sujan K. Gonugondla, Ameya Patil, and Naresh R. Shanbhag. A multifunctional in-memory inference processor using a standard 6T SRAM array. *IEEE Journal of Solid-State Circuits*, 53(2):642–655, 2018. DOI: 10.1109/jssc.2017.2782087 37, 39, 40, 41, 42, 43, 44, 62, 65

[67] Jintao Zhang, Zhuo Wang, and Naveen Verma. A machine-learning classifier implemented in a standard 6T SRAM array. In *IEEE Symposium on VLSI Circuits (VLSI-Circuits)*, pages 1–2, 2016. DOI: 10.1109/vlsic.2016.7573556 37, 38, 43, 62, 65

[68] Hossein Valavi, Peter J. Ramadge, Eric Nestler, and Naveen Verma. A mixed-signal binarized convolutional-neural-network accelerator integrating dense weight storage and multiplication for reduced data movement. In *IEEE Symposium on VLSI Circuits*, pages 141–142, 2018. DOI: 10.1109/vlsic.2018.8502421 39, 43, 62, 65

[69] Avishek Biswas and Anantha P. Chandrakasan. Conv-RAM: An energy-efficient SRAM with embedded convolution computation for low-power CNN-based machine learning applications. In *IEEE International Solid-State Circuits Conference-(ISSCC)*, pages 488–490, 2018. DOI: 10.1109/isscc.2018.8310397 39, 43, 62, 65

[70] Mingu Kang, Min-Sun Keel, Naresh R. Shanbhag, Sean Eilert, and Ken Curewitz. An energy-efficient VLSI architecture for pattern recognition via deep embedding of computation in SRAM. In *IEEE International Conference on Acoustics, Speech and Signal Processing (ICASSP)*, pages 8326–8330, 2014. DOI: 10.1109/icassp.2014.6855225 41

[71] Mingu Kang, Sujan K. Gonugondla, Min-Sun Keel, and Naresh R. Shanbhag. An energy-efficient memory-based high-throughput VLSI architecture for convolutional networks. In *IEEE International Conference on Acoustics, Speech and Signal Processing (ICASSP)*, pages 1037–1041, 2015. DOI: 10.1109/ICASSP.2015.7178127 41

[72] Kelin J. Kuhn. Reducing variation in advanced logic technologies: Approaches to process and design for manufacturability of nanoscale CMOS. In *IEEE International Electron Devices Meeting*, pages 471–474, 2007. DOI: 10.1109/IEDM.2007.4418976 41

[73] Prakalp Srivastava, Mingu Kang, Sujan K. Gonugondla, Sungmin Lim, Jungwook Choi, Vikram Adve, Nam Sung Kim, and Naresh Shanbhag. Promise: An end-to-end design of a programmable mixed-signal accelerator for machine-learning algorithms. In *ACM/IEEE 45th Annual International Symposium on Computer Architecture (ISCA)*, pages 43–56, 2018. DOI: 10.1109/isca.2018.00015 42, 43, 44, 62, 65

[74] Yifan Yuan, Yipeng Wang, Ren Wang, and Jian Huang. Halo: Accelerating flow classification for scalable packet processing in NFV. In *ACM/IEEE 46th Annual International Symposium on Computer Architecture (ISCA)*, pages 601–614, 2019. DOI: 10.1145/3307650.3322272 44

[75] Elliot Lockerman, Axel Feldmann, Mohammad Bakhshalipour, Alexandru Stanescu, Shashwat Gupta, Daniel Sanchez, and Nathan Beckmann. Livia: Data-centric computing throughout the memory hierarchy. In *Proc. of the 25th International Conference on Architectural Support for Programming Languages and Operating Systems*, pages 417–433, 2020. DOI: 10.1145/3373376.3378497 44

[76] Snehasish Kumar, Naveen Vedula, Arrvindh Shriraman, and Vijayalakshmi Srini-vasan. DASX: Hardware accelerator for software data structures. In *Proc. of the 29th ACM on International Conference on Supercomputing*, pages 361–372, 2015. DOI: 10.1145/2751205.2751231 44

[77] Ping Chi, Shuangchen Li, Cong Xu, Tao Zhang, Jishen Zhao, Yongpan Liu, Yu Wang, and Yuan Xie. Prime: A novel processing-in-memory architecture for neural network computation in ReRAM-based main memory. *ACM SIGARCH Computer Architecture News*, 44(3):27–39, 2016. DOI: 10.1145/3007787.3001140 46, 48, 49, 62

[78] Ali Shafiee, Anirban Nag, Naveen Muralimanohar, Rajeev Balasubramonian, John Paul Strachan, Miao Hu, R. Stanley Williams, and Vivek Srikumar. Isaac: A convolutional neural network accelerator with in-situ analog arithmetic in crossbars. *ACM SIGARCH Computer Architecture News*, 44(3):14–26, 2016. DOI: 10.1145/3007787.3001139 46, 48, 49, 51, 62, 63

[79] Boxun Li, Yi Shan, Miao Hu, Yu Wang, Yiran Chen, and Huazhong Yang. Memristor-based approximated computation. In *International Symposium on Low Power Electronics and Design (ISLPED)*, pages 242–247, 2013. DOI: 10.1109/ISLPED.2013.6629302 48

[80] Mirko Prezioso, Farnood Merrikh-Bayat, B. D. Hoskins, Gina C. Adam, Konstantin K. Likharev, and Dmitri B. Strukov. Training and operation of an integrated neuromor-phic network based on metal-oxide memristors. *Nature*, 521(7550):61–64, 2015. DOI: 10.1038/nature14441 48

[81] Yongtae Kim, Yong Zhang, and Peng Li. A reconfigurable digital neuromorphic processor with memristive synaptic crossbar for cognitive computing. *ACM Journal on Emerging Technologies in Computing Systems (JETC)*, 11(4):1–25, 2015. DOI: 10.1145/2700234 48

[82] M. Hu, J. P. Strachan, E. Merced-Grafals, Z. Li, and R. S. Williams. Dot-product en-gine: Programming memristor crossbar arrays for efficient vector-matrix multiplication. In *ICCAD'15 Workshop on Towards Efficient Computing in the Dark Silicon Era*, 2015. 48

[83] Shimeng Yu, Zhiwei Li, Pai-Yu Chen, Huaqiang Wu, Bin Gao, Deli Wang, Wei Wu, and He Qian. Binary neural network with 16 mb RRAM macro chip for classification and online training. In *IEEE International Electron Devices Meeting (IEDM)*, pages 16–2, 2016. DOI: 10.1109/iedm.2016.7838429 49

[84] Lixue Xia, Tianqi Tang, Wenqin Huangfu, Ming Cheng, Xiling Yin, Boxun Li, Yu Wang, and Huazhong Yang. Switched by input: Power efficient structure for RRAM-based convolutional neural network. In *Proc. of the 53rd Annual Design Automation Conference*, pages 1–6, 2016. DOI: 10.1145/2897937.2898101 49

[85] Deepak Kadetotad, Zihan Xu, Abinash Mohanty, Pai-Yu Chen, Binbin Lin, Jieping Ye, Sarma Vrudhula, Shimeng Yu, Yu Cao, and Jae-Sun Seo. Parallel architecture with resistive crosspoint array for dictionary learning acceleration. *IEEE Journal on Emerging and Selected Topics in Circuits and Systems*, 5(2):194–204, 2015. DOI: 10.1109/jetcas.2015.2426495 49

[86] Wei-Hao Chen, Kai-Xiang Li, Wei-Yu Lin, Kuo-Hsiang Hsu, Pin-Yi Li, Cheng-Han Yang, Cheng-Xin Xue, En-Yu Yang, Yen-Kai Chen, Yun-Sheng Chang, et al. A 65 nm 1 mb nonvolatile computing-in-memory ReRAM macro with sub-16 ns multiply-and-accumulate for binary DNN AI edge processors. In *IEEE International Solid–State Circuits Conference-(ISSCC)*, pages 494–496, 2018. DOI: 10.1109/isscc.2018.8310400 49, 62, 64

[87] Cheng-Xin Xue, Wei-Hao Chen, Je-Syu Liu, Jia-Fang Li, Wei-Yu Lin, Wei-En Lin, Jing-Hong Wang, Wei-Chen Wei, Ting-Wei Chang, Tung-Cheng Chang, et al. 24.1 a 1 mb multibit ReRAM computing-in-memory macro with 14.6 ns parallel MAC computing time for CNN-based AI edge processors. In *IEEE International Solid–State Circuits Conference-(ISSCC)*, pages 388–390, 2019. DOI: 10.1109/isscc.2019.8662395 49, 62, 64

[88] Shuangchen Li, Cong Xu, Qiaosha Zou, Jishen Zhao, Yu Lu, and Yuan Xie. Pinatubo: A processing-in-memory architecture for bulk bitwise operations in emerging non-volatile memories. In *Proc. of the 53rd Annual Design Automation Conference*, pages 1–6, 2016. DOI: 10.1145/2897937.2898064 49, 50, 58

[89] Mohsen Imani, Saransh Gupta, and Tajana Rosing. Ultra-efficient processing in-memory for data intensive applications. In *54th ACM/EDAC/IEEE Design Automation Conference (DAC)*, pages 1–6, 2017. DOI: 10.1145/3061639.3062337 50

[90] Ben Feinberg, Uday Kumar Reddy Vengalam, Nathan Whitehair, Shibo Wang, and Engin Ipek. Enabling scientific computing on memristive accelerators. In *ACM/IEEE 45th Annual International Symposium on Computer Architecture (ISCA)*, pages 367–382, 2018. DOI: 10.1109/isca.2018.00039 51

[91] Mohsen Imani, Saransh Gupta, Yeseong Kim, and Tajana Rosing. FloatPIM: In-memory acceleration of deep neural network training with high precision. In *ACM/IEEE 46th Annual International Symposium on Computer Architecture (ISCA)*, pages 802–815, 2019. DOI: 10.1145/3307650.3322237 51, 62, 64

[92] Linghao Song, Xuehai Qian, Hai Li, and Yiran Chen. PipeLayer: A pipelined ReRAM-based accelerator for deep learning. In *IEEE International Symposium on High Performance Computer Architecture (HPCA)*, pages 541–552, 2017. DOI: 10.1109/hpca.2017.55 51, 62, 64

[93] Teyuh Chou, Wei Tang, Jacob Botimer, and Zhengya Zhang. Cascade: Connecting RRAMs to extend analog dataflow in an end-to-end in-memory processing paradigm.

In *Proc. of the 52nd Annual IEEE/ACM International Symposium on Microarchitecture*, pages 114–125, 2019. DOI: 10.1145/3352460.3358328 51, 62

[94] Julien Borghetti, Gregory S. Snider, Philip J. Kuekes, J. Joshua Yang, Duncan R. Stewart, and R. Stanley Williams. "Memristive" switches enable "stateful" logic operations via material implication. *Nature*, 464(7290):873–876, 2010. DOI: 10.1038/nature08940 52

[95] Eike Linn, R. Rosezin, Stefan Tappertzhofen, U. Böttger, and Rainer Waser. Beyond Von Neumann—logic operations in passive crossbar arrays alongside memory operations. *Nanotechnology*, 23(30):305205, 2012. DOI: 10.1088/0957-4484/23/30/305205 52

[96] Pierre-Emmanuel Gaillardon, Luca Amarú, Anne Siemon, Eike Linn, Rainer Waser, Anupam Chattopadhyay, and Giovanni De Micheli. The programmable logic-in-memory (PLIM) computer. In *Design, Automation and Test in Europe Conference and Exhibition (DATE)*, pages 427–432, IEEE, 2016. DOI: 10.3850/9783981537079_0970 52

[97] Mathias Soeken, Saeideh Shirinzadeh, Pierre-Emmanuel Gaillardon, Luca Gaetano Amarú, Rolf Drechsler, and Giovanni De Micheli. An mig-based compiler for programmable logic-in-memory architectures. In *53nd ACM/EDAC/IEEE Design Automation Conference (DAC)*, pages 1–6, 2016. DOI: 10.1145/2897937.2897985 53

[98] Debjyoti Bhattacharjee, Rajeswari Devadoss, and Anupam Chattopadhyay. ReVAMP: ReRAM-based VLIW architecture for in-memory computing. In *Design, Automation and Test in Europe Conference and Exhibition (DATE)*, pages 782–787, IEEE, 2017. DOI: 10.23919/date.2017.7927095 53

[99] Anne Siemon, Stephan Menzel, Rainer Waser, and Eike Linn. A complementary resistive switch-based crossbar array adder. *IEEE Journal on Emerging and Selected Topics in Circuits and Systems*, 5(1):64–74, 2015. DOI: 10.1109/jetcas.2015.2398217 53

[100] Muath Abu Lebdeh, Heba Abunahla, Baker Mohammad, and Mahmoud Al-Qutayri. An efficient heterogeneous memristive XNOR for in-memory computing. *IEEE Transactions on Circuits and Systems I: Regular Papers*, 64(9):2427–2437, 2017. DOI: 10.1109/tcsi.2017.2706299 53

[101] Stewart A. Schuster, H. B. Nguyen, Esen A. Ozkarahan, and Kenneth C. Smith. RAP.2—an associative processor for databases and its applications. *IEEE Transactions on Computers*, 28(6):446–458, 1979. DOI: 10.1109/tc.1979.1675383 56

[102] Chyuan Shiun Lin, Diane C. P. Smith, and John Miles Smith. The design of a rotating associative memory for relational database applications. *ACM Transactions on Database Systems (TODS)*, 1(1):53–65, 1976. DOI: 10.1145/320434.320447 56

[103] Hans-Otto Leilich, Günther Stiege, and Hans Christoph Zeidler. A search processor for data base management systems. In *Proc. of the 4th International Conference on Very Large Data Bases—Volume 4*, pages 280–287, 1978. 56

[104] Ali R. Hurson, Les L. Miller, Simin H. Pakzad, Margaret H. Eich, and Behrooz Shirazi. Parallel architectures for database systems. *Advances in Computers*, 28:107–151, 1989. DOI: 10.1016/s0065-2458(08)60047-9 56

[105] Anurag Acharya, Mustafa Uysal, and Joel Saltz. Active disks: Programming model, algorithms and evaluation. *ACM SIGOPS Operating Systems Review*, 32(5):81–91, 1998. DOI: 10.1145/384265.291026 56

[106] Kimberly Keeton, David A. Patterson, and Joseph M. Hellerstein. A case for intelligent disks (IDISKs). *ACM SIGMOD Record*, 27(3):42–52, 1998. DOI: 10.1145/290593.290602 56

[107] Devesh Tiwari, Sudharshan S. Vazhkudai, Youngjae Kim, Xiaosong Ma, Simona Boboila, and Peter J. Desnoyers. Reducing data movement costs using energy-efficient, active computation on {SSD}. In *Workshop on Power-Aware Computing and Systems (Hot-Power 12)*, 2012. 56

[108] Yu-Ching Hu, Murtuza Taher Lokhandwala, Te I, and Hung-Wei Tseng. Dynamic multi-resolution data storage. In *Proc. of the 52nd Annual IEEE/ACM International Symposium on Microarchitecture*, pages 196–210, 2019. DOI: 10.1145/3352460.3358282 56

[109] Jaeyoung Do, Yang-Suk Kee, Jignesh M. Patel, Chanik Park, Kwanghyun Park, and David J. DeWitt. Query processing on smart SSDs: Opportunities and challenges. In *Proc. of the ACM SIGMOD International Conference on Management of Data*, pages 1221–1230, 2013. DOI: 10.1145/2463676.2465295 56, 78

[110] Louis Woods, Zsolt István, and Gustavo Alonso. Ibex: An intelligent storage engine with support for advanced SQL offloading. *Proc. of the VLDB Endowment*, 7(11):963–974, 2014. DOI: 10.14778/2732967.2732972 56, 78

[111] Adrian M. Caulfield, Arup De, Joel Coburn, Todor I. Mollow, Rajesh K. Gupta, and Steven Swanson. Moneta: A high-performance storage array architecture for next-generation, non-volatile memories. In *43rd Annual IEEE/ACM International Symposium on Microarchitecture*, pages 385–395, 2010. DOI: 10.1109/micro.2010.33 56

[112] Sudharsan Seshadri, Mark Gahagan, Sundaram Bhaskaran, Trevor Bunker, Arup De, Yanqin Jin, Yang Liu, and Steven Swanson. Willow: A user-programmable {SSD}. In *11th {USENIX} Symposium on Operating Systems Design and Implementation ({OSDI} 14)*, pages 67–80, 2014. 56, 78

[113] Mohit Saxena, Michael M. Swift, and Yiying Zhang. FlashTier: A lightweight, consistent and durable storage cache. In *Proc. of the 7th ACM European Conference on Computer Systems*, pages 267–280, 2012. DOI: 10.1145/2168836.2168863 56

[114] Yanqin Jin, Hung-Wei Tseng, Yannis Papakonstantinou, and Steven Swanson. KAML: A flexible, high-performance key-value SSD. In *IEEE International Symposium on High Performance Computer Architecture (HPCA)*, pages 373–384, 2017. DOI: 10.1109/hpca.2017.15 56

[115] Hung-Wei Tseng, Qianchen Zhao, Yuxiao Zhou, Mark Gahagan, and Steven Swanson. Morpheus: Creating application objects efficiently for heterogeneous computing. *ACM SIGARCH Computer Architecture News*, 44(3):53–65, 2016. DOI: 10.1145/3007787.3001143 56

[116] Sang-Woo Jun, Ming Liu, Sungjin Lee, Jamey Hicks, John Ankcorn, Myron King, Shuotao Xu, et al. BlueDBM: An appliance for big data analytics. In *ACM/IEEE 42nd Annual International Symposium on Computer Architecture (ISCA)*, pages 1–13, 2015. DOI: 10.1145/2749469.2750412 56

[117] Erik Riedel, Christos Faloutsos, Garth A. Gibson, and David Nagle. Active disks for large-scale data processing. *Computer*, 34(6):68–74, 2001. DOI: 10.1109/2.928624 56

[118] Yangwook Kang, Yang-suk Kee, Ethan L. Miller, and Chanik Park. Enabling cost-effective data processing with smart SSD. In *IEEE 29th Symposium on Mass Storage Systems and Technologies (MSST)*, pages 1–12, 2013. DOI: 10.1109/msst.2013.6558444 56

[119] I. Stephen Choi and Yang-Suk Kee. Energy efficient scale-in clusters with in-storage processing for big-data analytics. In *Proc. of the International Symposium on Memory Systems*, pages 265–273, 2015. DOI: 10.1145/2818950.2818983 56

[120] Gunjae Koo, Kiran Kumar Matam, I. Te, H. V. Krishna Giri Narra, Jing Li, Hung-Wei Tseng, Steven Swanson, and Murali Annavaram. Summarizer: Trading communication with computing near storage. In *50th Annual IEEE/ACM International Symposium on Microarchitecture (MICRO)*, pages 219–231, 2017. DOI: 10.1145/3123939.3124553 56

[121] Samsung smartSSD computational storage drive. https://web.archive.org/web/20210118224545/https://samsungsemiconductor-us.com/smartssd/index.html 56

[122] Scaleflux CSD2000 product brief. https://web.archive.org/web/20200805141851/https://www.scaleflux.com/downloads/[200204]ScaleFlux_Product_Brief_CSD2000.pdf 56

[123] Innogrit—startup puts AI core in SSDs. https://web.archive.org/web/20210326215656/http://innogritcorp.com/en/site/newsdetail/125 56

[124] Panni Wang, Feng Xu, Bo Wang, Bin Gao, Huaqiang Wu, He Qian, and Shimeng Yu. Three-dimensional NAND flash for vector—matrix multiplication. *IEEE Transactions on Very Large Scale Integration (VLSI) Systems*, 27(4):988–991, 2019. DOI: 10.1109/tvlsi.2018.2882194 57

[125] Mike Demler. Mythic multiplies in a flash. *Analog in-Memory Computing Eliminates DRAM Read/Write Cylcles, The Linley Group Microprocessor report*, 2018. 57

[126] Xinjie Guo, F. Merrikh Bayat, M. Bavandpour, M. Klachko, M. R. Mahmoodi, M. Prezioso, K. K. Likharev, and D. B. Strukov. Fast, energy-efficient, robust, and reproducible mixed-signal neuromorphic classifier based on embedded nor flash memory technology. In *IEEE International Electron Devices Meeting (IEDM)*, pages 6.5.1–6.5.4, 2017. DOI: 10.1109/iedm.2017.8268341 57

[127] Wang Kang, Haotian Wang, Zhaohao Wang, Youguang Zhang, and Weisheng Zhao. In-memory processing paradigm for bitwise logic operations in STT–MRAM. *IEEE Transactions on Magnetics*, 53(11):1–4, 2017. DOI: 10.1109/tmag.2017.2703863 58

[128] Farhana Parveen, Zhezhi He, Shaahin Angizi, and Deliang Fan. HieLM: Highly flexible in-memory computing using STT MRAM. In *23rd Asia and South Pacific Design Automation Conference (ASP-DAC)*, pages 361–366, IEEE, 2018. DOI: 10.1109/aspdac.2018.8297350 58

[129] Yu Pan, Peng Ouyang, Yinglin Zhao, Wang Kang, Shouyi Yin, Youguang Zhang, Weisheng Zhao, and Shaojun Wei. A multilevel cell STT-MRAM-based computing in-memory accelerator for binary convolutional neural network. *IEEE Transactions on Magnetics*, 54(11):1–5, 2018. DOI: 10.1109/tmag.2018.2848625 58

[130] Shubham Jain, Ashish Ranjan, Kaushik Roy, and Anand Raghunathan. Computing in memory with spin-transfer torque magnetic RAM. *IEEE Transactions on Very Large Scale Integration (VLSI) Systems*, 26(3):470–483, 2017. DOI: 10.1109/tvlsi.2017.2776954 58

[131] Mohammad Rastegari, Vicente Ordonez, Joseph Redmon, and Ali Farhadi. XNOR-Net: ImageNet classification using binary convolutional neural networks. In *European Conference on Computer Vision*, pages 525–542, Springer, 2016. DOI: 10.1007/978-3-319-46493-0_32 58

[132] Marco Cassinerio, N. Ciocchini, and Daniele Ielmini. Logic computation in phase change materials by threshold and memory switching. *Advanced Materials*, 25(41):5975–5980, 2013. DOI: 10.1002/adma.201301940 58

[133] C. David Wright, Peiman Hosseini, and Jorge A. Vazquez Diosdado. Beyond Von-Neumann computing with nanoscale phase-change memory devices. *Advanced Functional Materials*, 23(18):2248–2254, 2013. DOI: 10.1002/adfm.201202383 58

[134] C. David Wright, Yanwei Liu, Krisztian I. Kohary, Mustafa M. Aziz, and Robert J. Hicken. Arithmetic and biologically-inspired computing using phase-change materials. *Advanced Materials*, 23(30):3408–3413, 2011. DOI: 10.1002/adma.201101060 58

[135] Geoffrey W. Burr, Robert M. Shelby, Severin Sidler, Carmelo Di Nolfo, Junwoo Jang, Irem Boybat, Rohit S. Shenoy, Pritish Narayanan, Kumar Virwani, Emanuele U. Giacometti, et al. Experimental demonstration and tolerancing of a large-scale neural network (165,000 synapses) using phase-change memory as the synaptic weight element. *IEEE Transactions on Electron Devices*, 62(11):3498–3507, 2015. DOI: 10.1109/ted.2015.2439635 58

[136] G. W. Burr, P. Narayanan, R. M. Shelby, Severin Sidler, Irem Boybat, Carmelo di Nolfo, and Yusuf Leblebici. Large-scale neural networks implemented with non-volatile memory as the synaptic weight element: Comparative performance analysis (accuracy, speed, and power). In *IEEE International Electron Devices Meeting (IEDM)*, page 4, 2015. DOI: 10.1109/iedm.2015.7409625 58

[137] Abu Sebastian, Tomas Tuma, Nikolaos Papandreou, Manuel Le Gallo, Lukas Kull, Thomas Parnell, and Evangelos Eleftheriou. Temporal correlation detection using computational phase-change memory. *Nature Communications*, 8(1):1–10, 2017. DOI: 10.1038/s41467-017-01481-9 58

[138] M. Le Gallo, A. Sebastian, G. Cherubini, H. Giefers, and E. Eleftheriou. Compressed sensing recovery using computational memory. In *IEEE International Electron Devices Meeting (IEDM)*, pages 28–3, 2017. DOI: 10.1109/iedm.2017.8268469 58

[139] SR Nandakumar, Manuel Le Gallo, Irem Boybat, Bipin Rajendran, Abu Sebastian, and Evangelos Eleftheriou. Mixed-precision architecture based on computational memory for training deep neural networks. In *IEEE International Symposium on Circuits and Systems (ISCAS)*, pages 1–5, 2018. DOI: 10.1109/iscas.2018.8351656 58

[140] Geoffrey W. Burr, Matthew J. Breitwisch, Michele Franceschini, Davide Garetto, Kailash Gopalakrishnan, Bryan Jackson, Bülent Kurdi, Chung Lam, Luis A. Lastras, Alvaro Padilla, et al. Phase change memory technology. *Journal of Vacuum Science and Technology B, Nanotechnology and Microelectronics: Materials, Processing, Measurement, and Phenomena*, 28(2):223–262, 2010. DOI: 10.1116/1.3301579 59

[141] Rajeev Balasubramanian. Innovations in the memory system, In *Synthesis Lectures on Computer Architecture*, 14(2):1–151, Morgan & Claypool Publishers, 2019. 61

[142] Mahdi Nazm Bojnordi and Engin Ipek. Memristive Boltzmann machine: A hardware accelerator for combinatorial optimization and deep learning. In *IEEE International Symposium on High Performance Computer Architecture (HPCA)*, pages 1–13, 2016. DOI: 10.1109/hpca.2016.7446049 62

[143] Haiyu Mao, Mingcong Song, Tao Li, Yuting Dai, and Jiwu Shu. LerGAN: A zero-free, low data movement and PIM-based GAN architecture. In *51st Annual IEEE/ACM International Symposium on Microarchitecture (MICRO)*, pages 669–681, 2018. DOI: 10.1109/micro.2018.00060 62

[144] Tzu-Hsien Yang, Hsiang-Yun Cheng, Chia-Lin Yang, I-Ching Tseng, Han-Wen Hu, Hung-Sheng Chang, and Hsiang-Pang Li. Sparse ReRAM engine: Joint exploration of activation and weight sparsity in compressed neural networks. In *Proc. of the 46th International Symposium on Computer Architecture*, pages 236–249, 2019. DOI: 10.1145/3307650.3322271 62

[145] Ming Cheng, Lixue Xia, Zhenhua Zhu, Yi Cai, Yuan Xie, Yu Wang, and Huazhong Yang. Time: A training-in-memory architecture for RRAM-based deep neural networks. *IEEE Transactions on Computer-Aided Design of Integrated Circuits and Systems*, 38(5):834–847, 2018. DOI: 10.1109/tcad.2018.2824304 62

[146] Aayush Ankit, Izzat El Hajj, Sai Rahul Chalamalasetti, Geoffrey Ndu, Martin Foltin, R. Stanley Williams, Paolo Faraboschi, Wen-Mei W. Hwu, John Paul Strachan, Kaushik Roy, et al. Puma: A programmable ultra-efficient memristor-based accelerator for machine learning inference. In *Proc. of the 24th International Conference on Architectural Support for Programming Languages and Operating Systems*, pages 715–731, 2019. DOI: 10.1145/3297858.3304049 62, 86

[147] Yu Ji, Youyang Zhang, Xinfeng Xie, Shuangchen Li, Peiqi Wang, Xing Hu, Youhui Zhang, and Yuan Xie. FPSA: A full system stack solution for reconfigurable ReRAM-based NN accelerator architecture. In *Proc. of the 24th International Conference on Architectural Support for Programming Languages and Operating Systems*, pages 733–747, 2019. DOI: 10.1145/3297858.3304048 62, 64

[148] Stefano Ambrogio, Pritish Narayanan, Hsinyu Tsai, Robert M. Shelby, Irem Boybat, Carmelo di Nolfo, Severin Sidler, Massimo Giordano, Martina Bodini, Nathan C. P. Farinha, et al. Equivalent-accuracy accelerated neural-network training using analogue memory. *Nature*, 558(7708):60–67, 2018. DOI: 10.1038/s41586-018-0180-5 62, 64

[149] Shubham Jain, Ashish Ranjan, Kaushik Roy, and Anand Raghunathan. Computing in memory with spin-transfer torque magnetic RAM. *IEEE Transactions on Very Large Scale Integration (VLSI) Systems*, 26(3):470–483, 2017. DOI: 10.1109/tvlsi.2017.2776954 62, 64

[150] Yandong Luo, Panni Wang, Xiaochen Peng, Xiaoyu Sun, and Shimeng Yu. Benchmark of ferroelectric transistor-based hybrid precision synapse for neural network accelerator. *IEEE Journal on Exploratory Solid-State Computational Devices and Circuits*, 5(2):142–150, 2019. DOI: 10.1109/jxcdc.2019.2925061 62, 64

[151] Sujan Kumar Gonugondla, Mingu Kang, and Naresh Shanbhag. A 42pJ/decision 3.12TOPS/W robust in-memory machine learning classifier with on-chip training. In *IEEE International Solid-State Circuits Conference-(ISSCC)*, pages 490–492, 2018. DOI: 10.1109/isscc.2018.8310398 62, 65

[152] Jun Yang, Yuyao Kong, Zhen Wang, Yan Liu, Bo Wang, Shouyi Yin, and Longxin Shi. 24.4 sandwich-RAM: An energy-efficient in-memory BWN architecture with pulse-width modulation. In *IEEE International Solid-State Circuits Conference-(ISSCC)*, pages 394–396, 2019. DOI: 10.1109/isscc.2019.8662435 62, 65

[153] Mingu Kang, Sungmin Lim, Sujan Gonugondla, and Naresh R. Shanbhag. An in-memory VLSI architecture for convolutional neural networks. *IEEE Journal on Emerging and Selected Topics in Circuits and Systems*, 8(3):494–505, 2018. DOI: 10.1109/jet-cas.2018.2829522 62, 65

[154] Youngeun Kwon, Yunjae Lee, and Minsoo Rhu. TensorDIMM: A practical near-memory processing architecture for embeddings and tensor operations in deep learning. In *Proc. of the 52nd Annual IEEE/ACM International Symposium on Microarchitecture*, pages 740–753, 2019. DOI: 10.1145/3352460.3358284 62, 65

[155] Jiawen Liu, Hengyu Zhao, Matheus A. Ogleari, Dong Li, and Jishen Zhao. Processing-in-memory for energy-efficient neural network training: A heterogeneous approach. In *51st Annual IEEE/ACM International Symposium on Microarchitecture (MICRO)*, pages 655–668, 2018. DOI: 10.1109/micro.2018.00059 62, 65, 87

[156] Hyeonuk Kim, Jaehyeong Sim, Yeongjae Choi, and Lee-Sup Kim. NAND-Net: Minimizing computational complexity of in-memory processing for binary neural networks. In *IEEE International Symposium on High Performance Computer Architecture (HPCA)*, pages 661–673, 2019. DOI: 10.1109/hpca.2019.00017 62, 65

[157] Mohsen Imani, Saikishan Pampana, Saransh Gupta, Minxuan Zhou, Yeseong Kim, and Tajana Rosing. Dual: Acceleration of clustering algorithms using digital-based processing in-memory. In *53rd Annual IEEE/ACM International Symposium on Microarchitecture (MICRO)*, pages 356–371, 2020. DOI: 10.1109/micro50266.2020.00039 62

[158] Vivienne Sze, Yu-Hsin Chen, Tien-Ju Yang, and Joel S. Emer. Efficient processing of deep neural networks. *Synthesis Lectures on Computer Architecture*, 15(2):1–341, 2020. DOI: 10.2200/s01004ed1v01y202004cac050 62

[159] Yann LeCun. The MNIST database of handwritten digits. http://yann.lecun.com/exdb/mnist/, 1998. 64

[160] Yu-Hsin Chen, Joel Emer, and Vivienne Sze. Eyeriss: A spatial architecture for energy-efficient dataflow for convolutional neural networks. *ACM SIGARCH Computer Architecture News*, 44(3):367–379, 2016. DOI: 10.1145/3007787.3001177 65

[161] Christopher Grant Jones, Rose Liu, Leo Meyerovich, Krste Asanovic, and Rastislav Bodik. Parallelizing the web browser. In *Proc. of the 1st USENIX Workshop on Hot Topics in Parallelism*, 2009. 66

[162] Qiong Wang, Mohamed El-Hadedy, Ke Wang, and Kevin Skadron. Accelerating weeder: A DNA motif search tool using the micron automata processor. 2016. DOI: 10.1587/transinf.2017edp7051 66

[163] Krste Asanovic, Ras Bodik, Bryan Christopher Catanzaro, Joseph James Gebis, Parry Husbands, Kurt Keutzer, David A. Patterson, William Lester Plishker, John Shalf, Samuel Webb Williams, et al. The landscape of parallel computing research: A view from Berkeley. 2006. 66

[164] Zhijia Zhao, Bo Wu, and Xipeng Shen. Challenging the "embarrassingly sequential": Parallelizing finite state machine-based computations through principled speculation. In *Architectural Support for Programming Languages and Operating Systems, ASPLOS'14*, pages 543–558, Salt Lake City, UT, March 1–5, 2014. DOI: 10.1145/2541940.2541989 66

[165] Pascal Caron and Djelloul Ziadi. Characterization of Glushkov automata. *Theoretical Computer Science*, 233(1–2):75–90, 2000. DOI: 10.1016/s0304-3975(97)00296-x 68

[166] Chunkun Bo, Ke Wang, Jeffrey J. Fox, and Kevin Skadron. Entity resolution acceleration using micron's automata processor. *Architectures and Systems for Big Data (ASBD), in Conjunction with ISCA*, 2015. 69

[167] Indranil Roy and Srinivas Aluru. Discovering motifs in biological sequences using the micron automata processor. *IEEE/ACM Transactions on Computational Biology and Bioinformatics*, 13(1):99–111, 2016. DOI: 10.1109/tcbb.2015.2430313 69, 73, 84

[168] Kevin Angstadt, Westley Weimer, and Kevin Skadron. Rapid programming of pattern-recognition processors. In *Proc. of the 21st International Conference on Architectural Support for Programming Languages and Operating Systems*, pages 593–605, ACM, 2016. DOI: 10.1145/2872362.2872393 72

[169] Kevin Angstadt, Jack Wadden, Westley Weimer, and Kevin Skadron. Portable programming with rapid. *IEEE Transactions on Parallel and Distributed Systems*, 30(4):939–952, 2018. DOI: 10.1109/tpds.2018.2869736 72

[170] Kevin Angstadt, Jack Wadden, Vinh Dang, Ted Xie, Dan Kramp, Westley Weimer, Mircea Stan, and Kevin Skadron.   MNCaRT: An open-source, multi-architecture automata-processing research and execution ecosystem. *IEEE Computer Architecture Letters*, 17(1):84–87, 2017. DOI: 10.1109/lca.2017.2780105 72

[171] Matthew Casias, Kevin Angstadt, Tommy Tracy II, Kevin Skadron, and Westley Weimer. Debugging support for pattern-matching languages and accelerators. In *Proc. of the 24th International Conference on Architectural Support for Programming Languages and Operating Systems*, pages 1073–1086, 2019. DOI: 10.1145/3297858.3304066 72

[172] Jack Wadden, Kevin Angstadt, and Kevin Skadron. Characterizing and mitigating output reporting bottlenecks in spatial automata processing architectures. In *IEEE International Symposium on High Performance Computer Architecture (HPCA)*, pages 749–761, 2018. DOI: 10.1109/hpca.2018.00069 72

[173] Hongyuan Liu, Mohamed Ibrahim, Onur Kayiran, Sreepathi Pai, and Adwait Jog. Architectural support for efficient large-scale automata processing. In *51st Annual IEEE/ACM International Symposium on Microarchitecture (MICRO)*, pages 908–920, 2018. DOI: 10.1109/micro.2018.00078 73

[174] Arun Subramaniyan and Reetuparna Das. Parallel automata processor. In *ACM/IEEE 44th Annual International Symposium on Computer Architecture (ISCA)*, pages 600–612, 2017. DOI: 10.1145/3079856.3080207 73

[175] Todd Mytkowicz, Madanlal Musuvathi, and Wolfram Schulte. Data-parallel finite-state machines. In *Proc. of the 19th International Conference on Architectural Support for Programming Languages and Operating Systems*, pages 529–542, 2014. DOI: 10.1145/2541940.2541988 73

[176] Sparsh Mittal.   A survey on applications and architectural-optimizations of micron's automata processor. *Journal of Systems Architecture*, 98:135–164, 2019. DOI: 10.1016/j.sysarc.2019.07.006 73

[177] Arun Subramaniyan, Jingcheng Wang, Ezhil R. M. Balasubramanian, David Blaauw, Dennis Sylvester, and Reetuparna Das. Cache automaton. In *Proc. of the 50th Annual IEEE/ACM International Symposium on Microarchitecture*, pages 259–272, 2017. DOI: 10.1145/3123939.3123986 73

[178] Elaheh Sadredini, Reza Rahimi, Vaibhav Verma, Mircea Stan, and Kevin Skadron. EAP: A scalable and efficient in-memory accelerator for automata processing. In *Proc. of the 52nd Annual IEEE/ACM International Symposium on Microarchitecture*, pages 87–99, 2019. DOI: 10.1145/3352460.3358324 73

[179] Jintao Yu, Hoang Anh Du Nguyen, Lei Xie, Mottaqiallah Taouil, and Said Ham-dioui. Memristive devices for computation-in-memory. In *Design, Automation and Test in Europe Conference and Exhibition (DATE)*, pages 1646–1651, IEEE, 2018. DOI: 10.23919/date.2018.8342278 73

[180] Elaheh Sadredini, Reza Rahimi, Marzieh Lenjani, Mircea Stan, and Kevin Skadron. Impala: Algorithm/architecture co-design for in-memory multi-stride pattern matching. In *IEEE International Symposium on High Performance Computer Architecture (HPCA)*, pages 86–98, 2020. DOI: 10.1109/hpca47549.2020.00017 73

[181] Elaheh Sadredini, Reza Rahimi, Marzieh Lenjani, Mircea Stan, and Kevin Skadron. FlexAmata: A universal and efficient adaption of applications to spatial automata pro-cessing accelerators. In *Proc. of the 25th International Conference on Architectural Sup-port for Programming Languages and Operating Systems*, pages 219–234, 2020. DOI: 10.1145/3373376.3378459 73

[182] Lifeng Nai, Ramyad Hadidi, Jaewoong Sim, Hyojong Kim, Pranith Kumar, and Hye-soon Kim. GraphPIM: Enabling instruction-level PIM offloading in graph computing frameworks. In *IEEE International Symposium on High Performance Computer Architecture (HPCA)*, pages 457–468, 2017. DOI: 10.1109/hpca.2017.54 74, 76, 86

[183] Mingxing Zhang, Youwei Zhuo, Chao Wang, Mingyu Gao, Yongwei Wu, Kang Chen, Christos Kozyrakis, and Xuehai Qian. GraphP: Reducing communication for PIM-based graph processing with efficient data partition. In *IEEE International Sympo-sium on High Performance Computer Architecture (HPCA)*, pages 544–557, 2018. DOI: 10.1109/hpca.2018.00053 75

[184] Youwei Zhuo, Chao Wang, Mingxing Zhang, Rui Wang, Dimin Niu, Yanzhi Wang, and Xuehai Qian. GraphQ: Scalable PIM-based graph processing. In *Proc. of the 52nd Annual IEEE/ACM International Symposium on Microarchitecture*, pages 712–725, 2019. DOI: 10.1145/3352460.3358256 75

[185] Guohao Dai, Tianhao Huang, Yuze Chi, Jishen Zhao, Guangyu Sun, Yongpan Liu, Yu Wang, Yuan Xie, and Huazhong Yang. GrapHH: A processing-in-memory architecture for large-scale graph processing. *IEEE Transactions on Computer-Aided Design of Integrated Circuits and Systems*, 38(4):640–653, 2018. DOI: 10.1109/tcad.2018.2821565 75

[186] Linghao Song, Youwei Zhuo, Xuehai Qian, Hai Li, and Yiran Chen. GraphR: Accelerat-ing graph processing using ReRAM. In *IEEE International Symposium on High Performance Computer Architecture (HPCA)*, pages 531–543, 2018. DOI: 10.1109/hpca.2018.00052 76

[187] Jian Xu and Steven Swanson. NOVA: A log-structured file system for hybrid volatile/non-volatile main memories. In *14th USENIX Conference on File and Storage*

Technologies (FAST 16), pages 323–338, Association, Santa Clara, CA, February 2016. https://www.usenix.org/conference/fast16/technical-sessions/presentation/xu 77

[188] Joseph Izraelevitz, Jian Yang, Lu Zhang, Juno Kim, Xiao Liu, Amirsaman Memaripour, Yun Joon Soh, Zixuan Wang, Yi Xu, Subramanya R. Dulloor, et al. Basic performance measurements of the intel optane DC persistent memory module. *ArXiv Preprint ArXiv:1903.05714*, 2019. 77

[189] Anthony Gutierrez, Michael Cieslak, Bharan Giridhar, Ronald G. Dreslinski, Luis Ceze, and Trevor Mudge. Integrated 3D-stacked server designs for increasing physical density of key-value stores. In *Proc. of the 19th International Conference on Architectural Support for Programming Languages and Operating Systems*, pages 485–498, 2014. DOI: 10.1145/2541940.2541951 78

[190] Arup De, Maya Gokhale, Rajesh Gupta, and Steven Swanson. Minerva: Accelerating data analysis in next-generation SSDs. In *IEEE 21st Annual International Symposium on Field-Programmable Custom Computing Machines*, pages 9–16, 2013. DOI: 10.1109/fccm.2013.46 78

[191] Eugene W. Myers and Webb Miller. Optimal alignments in linear space. *Bioinformatics*, 4(1):11–17, 1988. DOI: 10.1093/bioinformatics/4.1.11 81

[192] Florian Plaza Onate, Jean-Michel Batto, Catherine Juste, Jehane Fadlallah, Cyrielle Fougeroux, Doriane Gouas, Nicolas Pons, Sean Kennedy, Florence Levenez, Joel Dore, et al. Quality control of microbiota metagenomics by k-mer analysis. *BMC Genomics*, 16(1):1–10, 2015. DOI: 10.1186/s12864-015-1406-7 82

[193] Sergey Koren, Brian P. Walenz, Konstantin Berlin, Jason R. Miller, Nicholas H. Bergman, and Adam M. Phillippy. Canu: scalable and accurate long-read assembly via adaptive k-mer weighting and repeat separation. *Genome Research*, 27(5):722–736, 2017. DOI: 10.1101/gr.215087.116 82

[194] Wenqin Huangfu, Xueqi Li, Shuangchen Li, Xing Hu, Peng Gu, and Yuan Xie. Medal: Scalable DIMM-based near data processing accelerator for DNA seeding algorithm. In *Proc. of the 52nd Annual IEEE/ACM International Symposium on Microarchitecture*, pages 587–599, 2019. DOI: 10.1145/3352460.3358329 83

[195] Farzaneh Zokaee, Mingzhe Zhang, and Lei Jiang. Finder: Accelerating FM-index-based exact pattern matching in genomic sequences through ReRAM technology. In *28th International Conference on Parallel Architectures and Compilation Techniques (PACT)*, pages 284–295, IEEE, 2019. DOI: 10.1109/pact.2019.00030 83

[196] Jeremie S. Kim, Damla Senol Cali, Hongyi Xin, Donghyuk Lee, Saugata Ghose, Mohammed Alser, Hasan Hassan, Oguz Ergin, Can Alkan, and Onur Mutlu. Grim-filter:

Fast seed location filtering in DNA read mapping using processing-in-memory technologies. *BMC Genomics*, 19(2):23–40, 2018. DOI: 10.1186/s12864-018-4460-0 83

[197] Anirban Nag, C. N. Ramachandra, Rajeev Balasubramonian, Ryan Stutsman, Edouard Giacomin, Hari Kambalasubramanyam, and Pierre-Emmanuel Gaillardon. Gencache: Leveraging in-cache operators for efficient sequence alignment. In *Proc. of the 52nd Annual IEEE/ACM International Symposium on Microarchitecture*, pages 334–346, 2019. DOI: 10.1145/3352460.3358308 83

[198] Roman Kaplan, Leonid Yavits, and Ran Ginosar. RASSA: Resistive pre-alignment accelerator for approximate DNA long read mapping. *IEEE Micro*, 39(4):44–54, 2018. DOI: 10.1109/mm.2018.2890253 83

[199] Daichi Fujiki, Arun Subramaniyan, Tianjun Zhang, Yu Zeng, Reetuparna Das, David Blaauw, and Satish Narayanasamy. GenAx: A genome sequencing accelerator. In *ACM/IEEE 45th Annual International Symposium on Computer Architecture (ISCA)*, pages 69–82, 2018. DOI: 10.1109/isca.2018.00017 83

[200] Yatish Turakhia, Gill Bejerano, and William J. Dally. Darwin: A genomics co-processor provides up to 15,000 x acceleration on long read assembly. *ACM SIGPLAN Notices*, 53(2):199–213, 2018. DOI: 10.1145/3296957.3173193 83

[201] Roman Kaplan, Leonid Yavits, Ran Ginosar, and Uri Weiser. A resistive CAM processing-in-storage architecture for DNA sequence alignment. *IEEE Micro*, 37(4):20–28, 2017. DOI: 10.1109/mm.2017.3211121 83

[202] Roman Kaplan, Leonid Yavits, and Ran Ginosasr. BioSEAL: In-memory biological sequence alignment accelerator for large-scale genomic data. In *Proc. of the 13th ACM International Systems and Storage Conference*, pages 36–48, 2020. DOI: 10.1145/3383669.3398279 83

[203] Saransh Gupta, Mohsen Imani, Behnam Khaleghi, Venkatesh Kumar, and Tajana Rosing. Rapid: A ReRAM processing in-memory architecture for DNA sequence alignment. In *IEEE/ACM International Symposium on Low Power Electronics and Design (ISLPED)*, pages 1–6, 2019. DOI: 10.1109/islped.2019.8824830 83

[204] Wenqin Huangfu, Shuangchen Li, Xing Hu, and Yuan Xie. Radar: A 3D-ReRAM-based DNA alignment accelerator architecture. In *Proc. of the 55th Annual Design Automation Conference*, pages 1–6, 2018. DOI: 10.1145/3195970.3196098 83

[205] Nathaniel Mcvicar, Chih-Ching Lin, and Scott Hauck. K-Mer counting using bloom filters with an FPGA-attached HMC. In *IEEE 25th Annual International Symposium on Field-Programmable Custom Computing Machines (FCCM)*, pages 203–210, 2017. DOI: 10.1109/fccm.2017.23 84

[206] Biresh Kumar Joardar, Priyanka Ghosh, Partha Pratim Pande, Ananth Kalyanaraman, and Sriram Krishnamoorthy. NoC-enabled software/hardware co-design framework for accelerating k-mer counting. In *Proc. of the 13th IEEE/ACM International Symposium on Networks-on-Chip*, pages 1–8, 2019. DOI: 10.1145/3313231.3352367 84

[207] Wenqin Huangfu, Krishna T. Malladi, Shuangchen Li, Peng Gu, and Yuan Xie. Nest: DIMM-based near-data-processing accelerator for k-mer counting. In *IEEE/ACM International Conference on Computer Aided Design (ICCAD)*, pages 1–9, 2020. DOI: 10.1145/3400302.3415724 84

[208] Shaahin Angizi, Jiao Sun, Wei Zhang, and Deliang Fan. PIM-Aligner: A processing-in-MRAM platform for biological sequence alignment. In *Design, Automation and Test in Europe Conference and Exhibition (DATE)*, pages 1265–1270, IEEE, 2020. DOI: 10.23919/date48585.2020.9116303 84

[209] Shaahin Angizi, Jiao Sun, Wei Zhang, and Deliang Fan. Aligns: A processing-in-memory accelerator for DNA short read alignment leveraging sot-MRAM. In *56th ACM/IEEE Design Automation Conference (DAC)*, pages 1–6, 2019. DOI: 10.1145/3316781.3317764 84

[210] Shaahin Angizi, Naima Ahmed Fahmi, Wei Zhang, and Deliang Fan. PIM-Assembler: A processing-in-memory platform for genome assembly. In *57th ACM/IEEE Design Automation Conference (DAC)*, pages 1–6, 2020. DOI: 10.1109/dac18072.2020.9218653 84

[211] Qian Lou, Sarath Chandra Janga, and Lei Jiang. Helix: Algorithm/architecture co-design for accelerating nanopore genome base-calling. In *Proc. of the ACM International Conference on Parallel Architectures and Compilation Techniques, PACT'20*, pages 293–304, Association for Computing Machinery, New York, 2020. https://doi.org/10.1145/3410463.3414626 DOI: 10.1145/3410463.3414626 84

[212] Mark Oskin, Frederic T. Chong, and Timothy Sherwood. Active pages: A computation model for intelligent memory. In *Proc. 25th Annual International Symposium on Computer Architecture (Cat. no. 98CB36235)*, pages 192–203, IEEE, 1998. DOI: 10.1109/isca.1998.694774 85

[213] Mary Hall, Apoorv Srivastava, William Athas, Vincent Freeh, Jaewook Shin, Joonseok Park, Peter Kogge, Jeff Koller, Pedro Diniz, Jacqueline Chame, Jeff Draper, Jeff LaCoss, John Granacki, and Jay Brockman. Mapping irregular applications to DIVA, a PIM-based data-intensive architecture. In *Proc. of the ACM/IEEE Conference on Supercomputing (CDROM)—Supercomputing'99*, page 57–57, ACM Press, New York, 1999. http://portal.acm.org/citation.cfm?doid=331532.331589 DOI: 10.1145/331532.331589 85

[214] Jay B. Brockman, Peter M. Kogge, Thomas L. Sterling, Vincent W. Freeh, and Shannon K. Kuntz. Microservers: A new memory semantics for massively parallel computing. In *Proc. of the 13th International Conference on Supercomputing*, pages 454–463, 1999. DOI: 10.1145/305138.305234 85

[215] Thorsten Von Eicken, David E. Culler, Seth Copen Goldstein, and Klaus Erik Schauser. Active messages: A mechanism for integrated communication and computation. *ACM SIGARCH Computer Architecture News*, 20(2):256–266, 1992. DOI: 10.1145/146628.140382 85

[216] Peter M. Kogge. Of piglets and threadlets: Architectures for self-contained, mobile, memory programming. In *Innovative Architecture for Future Generation High-Performance Processors and Systems (IWIA'04)*, pages 130–138, IEEE, 2004. DOI: 10.1109/iwia.2004.10005 85

[217] Saugata Ghose, Amirali Boroumand, Jeremie S. Kim, Juan Gómez-Luna, and Onur Mutlu. Processing-in-memory: A workload-driven perspective. *IBM Journal of Research and Development*, 63(6):3:1–3:19, 2019. DOI: 10.1147/jrd.2019.2934048 86

[218] Onur Mutlu, Saugata Ghose, Juan Gómez-Luna, and Rachata Ausavarungnirun. A modern primer on processing in memory. *ArXiv Preprint ArXiv:2012.03112*, 2020. 86

[219] Saugata Ghose, Kevin Hsieh, Amirali Boroumand, Rachata Ausavarungnirun, and Onur Mutlu. Enabling the adoption of processing-in-memory: Challenges, mechanisms, future research directions. *ArXiv Preprint ArXiv:1802.00320*, 2018. 86

[220] Saugata Ghose, Amirali Boroumand, Jeremie S. Kim, Juan Gómez-Luna, and Onur Mutlu. Processing-in-memory: A workload-driven perspective. *IBM Journal of Research and Development*, 63(6):3–1, 2019. DOI: 10.1147/jrd.2019.2934048 86

[221] Ramyad Hadidi, Lifeng Nai, Hyojong Kim, and Hyesoon Kim. Cairo: A compiler-assisted technique for enabling instruction-level offloading of processing-in-memory. *ACM Transactions on Architecture and Code Optimization (TACO)*, 14(4):1–25, 2017. DOI: 10.1145/3155287 86

[222] Amirali Boroumand, Saugata Ghose, Youngsok Kim, Rachata Ausavarungnirun, Eric Shiu, Rahul Thakur, Daehyun Kim, Aki Kuusela, Allan Knies, Parthasarathy Ranganathan, et al. Google workloads for consumer devices: Mitigating data movement bottlenecks. In *Proc. of the 23rd International Conference on Architectural Support for Programming Languages and Operating Systems*, pages 316–331, 2018. DOI: 10.1145/3173162.3173177 86

[223] Ravi Nair, Samuel F. Antao, Carlo Bertolli, Pradip Bose, Jose R. Brunheroto, Tong Chen, C-Y Cher, Carlos H. A. Costa, Jun Doi, Constantinos Evangelinos, et al. Active memory

cube: A processing-in-memory architecture for exascale systems. *IBM Journal of Research and Development*, 59(2/3):17–1, 2015. DOI: 10.1147/jrd.2015.2409732 87

[224] Mohammad Alian, Seung Won Min, Hadi Asgharimoghaddam, Ashutosh Dhar, Dong Kai Wang, Thomas Roewer, Adam McPadden, Oliver O'Halloran, Deming Chen, Jinjun Xiong, et al. Application-transparent near-memory processing architecture with memory channel network. In *51st Annual IEEE/ACM International Symposium on Microarchitecture (MICRO)*, pages 802–814, 2018. DOI: 10.1109/micro.2018.00070 87

[225] Kevin Hsieh, Eiman Ebrahimi, Gwangsun Kim, Niladrish Chatterjee, Mike O'Connor, Nandita Vijaykumar, Onur Mutlu, and Stephen W. Keckler. Transparent offloading and mapping (TOM) enabling programmer-transparent near-data processing in GPU systems. *ACM SIGARCH Computer Architecture News*, 44(3):204–216, 2016. DOI: 10.1145/3007787.3001159 87

[226] Ashutosh Pattnaik, Xulong Tang, Onur Kayiran, Adwait Jog, Asit Mishra, Mahmut T. Kandemir, Anand Sivasubramaniam, and Chita R. Das. Opportunistic computing in GPU architectures. In *ACM/IEEE 46th Annual International Symposium on Computer Architecture (ISCA)*, pages 210–223, 2019. DOI: 10.1145/3307650.3322212 87

[227] Daichi Fujiki, Scott Mahlke, and Reetuparna Das. In-memory data parallel processor. *ACM SIGPLAN Notices*, 53(2):1–14, 2018. DOI: 10.1145/3296957.3173171 89, 91

[228] Martin Abadi, Ashish Agarwal, Paul Barham, Eugene Brevdo, Zhifeng Chen, Craig Citro, Greg Corrado, Andy Davis, Jeffrey Dean, Matthieu Devin, Sanjay Ghemawat, Ian Goodfellow, Andrew Harp, Geoffrey Irving, Michael Isard, Yangqing Jia, Lukasz Kaiser, Manjunath Kudlur, Josh Levenberg, Dan Man, Rajat Monga, Sherry Moore, Derek Murray, Jon Shlens, Benoit Steiner, Ilya Sutskever, Paul Tucker, Vincent Vanhoucke, Vijay Vasudevan, Oriol Vinyals, Pete Warden, Martin Wicke, Yuan Yu, and Xiaoqiang Zheng. TensorFlow: Large-scale machine learning on heterogeneous distributed systems. *ArXiv:1603.04467*, page 19, 2015. http://download.tensorflow.org/paper/whitepaper2015.pdf 89

[229] John R. Ellis. *Bulldog: A Compiler for VLSI Architectures*, MIT Press, Cambridge, MA, 1986. 93

[230] Henry Cook, Krste Asanovic, and David A. Patterson. Virtual local stores: Enabling software-managed memory hierarchies in mainstream computing environments. *Technical Report no. UCB/EECS-2009-131*, 2009. 94

[231] Rakesh Komuravelli, Matthew D. Sinclair, Johnathan Alsop, Muhammad Huzaifa, Maria Kotsifakou, Prakalp Srivastava, Sarita V. Adve, and Vikram S. Adve. Stash: Have your scratchpad and cache it too. *ACM SIGARCH Computer Architecture News*, 43(3S):707–719, 2015. DOI: 10.1145/2872887.2750374 94

[232] Peter Pessl, Daniel Gruss, Clémentine Maurice, Michael Schwarz, and Stefan Mangard. Drama: Exploiting DRAM addressing for cross-CPU attacks. In *25th USENIX security symposium (USENIX security 16)*, pages 565–581, 2016. 94

Authors' Biographies

DAICHI FUJIKI

Daichi Fujiki received his B.E. degree from Keio University, Tokyo, Japan, in 2016 and his M.S.Eng. degree from the University of Michigan, Ann Arbor, MI, in 2017. He is currently pursuing a Ph.D. in Computer Science and Engineering with the University of Michigan, Ann Arbor, MI. He is a member of the Mbits Research Group, Computer Engineering Laboratory (CELAB), University of Michigan, which develops in-situ compute memory architectures and custom acceleration hardware for bioinformatics workloads.

XIAOWEI WANG

Xiaowei Wang received his B.Eng. degree in Electronic Information Science and Technology from Tsinghua University, Beijing, China, in 2015. He received his M.S. degree in Computer Science and Engineering from the University of Michigan, Ann Arbor, MI, in 2017, where he is currently pursuing a Ph.D. in Computer Science and Engineering. He is advised by Prof. Reetuparna Das. His research interests include domain-specific architectures for machine learning, in-memory computing, and hardware/software co-design.

ARUN SUBRAMANIYAN

Arun Subramaniyan received his B.E (Hons.) in Electrical and Electronics from the Birla Institute of Technology and Science (BITS-Pilani), India in 2015. He is currently a Ph.D. student at the University of Michigan, advised by Prof. Reetuparna Das. His dissertation research focuses on developing efficient algorithms and customized computing systems for precision health. He is also interested in in-memory computing architectures and hardware reliability. His work has been recognized by UM's Precision Health Scholars Award, a Rackham International Students Fellowship, an IEEE Micro Top Picks Award, and a Best Paper Award in CODES+ISSS.

REETUPARNA DAS

Reetuparna Das received her Ph.D. in Computer Science and Engineering from the Pennsylvania State University, University Park, PA, in 2010. She was a Research Scientist with the Intel Labs, Santa Clara, CA, and the Researcher-In-Residence with the Center for Future Architectures Research, Ann Arbor, MI. She is currently an Associate Professor with the University of

Michigan, Ann Arbor. Some of her recent projects include in-memory architectures, custom computing for precision health and AI, fine-grain heterogeneous core architectures for mobile systems, and low-power scalable interconnects for kilo-core processors. She has authored over 45 articles and holds 7 patents. Prof. Das received two IEEE top picks awards, the NSF CAREER Award, the CRA-W's Borg Early Career Award, the Intel Outstanding Researcher Award, and the Sloan Foundation Fellowship. She has been inducted into IEEE/ACM MICRO and ISCA Hall of Fame. She served on over 30 technical program committees and as the Program Co-Chair for MICRO-52.

Printed in the United States
by Baker & Taylor Publisher Services